Wind Energy Landscape

IEEE Press
445 Hoes Lane
Piscataway, NJ 08854

Wind Energy Landscape

Principles, Techniques, History

James S. Aber and Susan E. W. Aber
Emporia State University, Kansas, USA

Firooza Pavri
University of Southern Maine, USA

Power & Energy Society®

IEEE Press

WILEY

Published by John Wiley & Sons, Inc., Hoboken, New Jersey.

Library of Congress Cataloging-in-Publication Data applied for

Hardback ISBN: 9781394212873

Cover Design: Wiley
Cover Image: © Imagery by the Authors

Set in 9.5/12.5pt STIXTwoText by Straive, Pondicherry, India

Contents

About the Authors

Susan E. W. Aber is a Halstead, Kansas native, and **James S. Aber** was born and raised in Kansas City, Missouri. Since meeting as students at the University of Kansas and marrying, they have resided most of their adult years in Emporia, Kansas. JSA received his undergraduate degree from Binghamton University, New York, and a PhD in Geology from the University of Kansas. He is a Distinguished Professor Emeritus at Emporia State University. His specialties include glaciation, wetlands, tectonics, and remote sensing. SEWA received her undergraduate degree in geology from the University of Kansas and a PhD in Library and Information Management from Emporia State University. She is faculty emerita at San José State University, California. She has expertise in mineralogy, gemstones, and maps and GIS for librarians. They have wide-ranging experiences and have conducted field research across the United States, Canada, and several countries in northern and central Europe. Both have conducted considerable research about general geology, glaciation, wetlands, and wind energy. They are partners for kite aerial photography and have published many scientific articles and several scholarly and popular books.

Firooza Pavri, PhD, is Professor of Geography and Associate Dean at the Edmund S. Muskie School of Public Service, University of Southern Maine (USM), USA. Prior to joining USM, she lived in the Midwest and received her MA in Geography & Planning from the University of Toledo and her PhD in Geography from the Ohio State University. Professor Pavri's teaching and research are in the field of environmental geography, with a focus on society–environment interactions, natural resource and energy policy, sustainable development, and geospatial technologies, including satellite imaging. Over the years, her research projects and field sites have included locations in the United States, Greenland, Iceland, and India.

Wind Energy Preface

The radical middle is where the solution to our energy challenges lies (Tinker 2013, p. 8).

Emergence of Wind Energy

People have harnessed the power in the wind since ancient times. Sailing, traditional windmills, and kites are just a few common examples found around the world. From the twelfth century until the early 1900s, in fact, traditional windmills provided significant motive power for grinding grain, pumping water, running small machines, and performing myriad mechanical tasks in many locales (Fig. P1). The Industrial Revolution stimulated development of new forms of energy derived mainly from fossil fuels along with widespread generation and use of electricity for diverse applications in modern society.

Coupled with rapid growth of human population, energy production and consumption increased even more quickly during the twentieth century in order to raise living standards and health for billions of people. Most people believed that cheap energy from coal, oil, gas, and uranium would fuel the future at low cost with trivial environmental consequences and minimal risks. While this situation appears naive in the twenty-first century, it seemed quite reasonable in the past. However, the era of low-cost energy is just a distant memory, environmental issues have come to the forefront, and risks of fossil and nuclear fuels are now recurring themes.

The use of wind power for generating electrical energy has grown phenomenally during the past half century from negligible in the 1970s to more than 1,000 gigawatts (GW) capacity today (GWEC 2024). With this rapid growth came recognition of the **windscape**, which is the combination of local climate and geography, environmental and ecological conditions, economic incentives and public policies, human land use and infrastructure, as well as historical and cultural expectations associated with harnessing wind power (Aber et al. 2015). Such an integrated

Figure P1 Traditional Dutch-style windmill at Trönningenäs, Halland, southwestern Sweden. The windmill dates from 1899; it has been fully restored and remains functional. When in operation, cloth sails cover the wooden frameworks of blades.

approach to assessing the wind-energy sector would allow for a full accounting of its costs and benefits and enable more efficient long-term planning strategies (Fig. P2).

Figure P2 Typical windscape on the High Plains in southwestern Kansas. Mixed agricultural land use and GE Wind 1.5 MW turbines in the original Spearville wind farm that went online in 2006. The nearest turbine stands less than 200 m from the farmstead (on left). Such close spacing is generally not allowed in newer wind farms. Kite aerial photograph with Unruh and Leiker (2008).

Role of Wind Energy

Wind energy has emerged from an uncertain niche industry to become an important component of national energy policies in many developed and developing countries. The case for increased wind power is based on several issues.

- Cost of fossil fuel—The supply and pricing of fossil fuels (oil, gas, coal) have been highly volatile since the 1970s. The long-term trend is for higher prices, but with wide market swings depending on slight imbalances in supply and demand.
- Dependence on foreign sources—Mass transfer of wealth to foreign sources of fuel is a severe economic drain, and uncertainty of supply for political or economic reasons jeopardizes national security for many developed and developing countries.
- Greenhouse gas emissions—Producing and burning fossil fuels emit large volumes of carbon dioxide and other greenhouse gases into the atmosphere. Coal is the worst offender in this regard, and coal-fired power plants are the major sources of electricity in many parts of the world.

  Homo sapiens are transforming the Earth through vast land conversions, immense use of minerals and fossil fuels, intense water consumption, and exploitation of many other non-renewable land and marine resources. This development has been costly for the natural environment. Human activity is largely uncontrolled; most decisions are made for immediate individual survival, gain, or comfort rather than the greater good of society or long-term environmental sustainability of the world. Nonetheless, increased generation and delivery of electricity are key ingredients for economic development and human quality of life.

  In the future, wind energy undoubtedly will play a critical role in the mix of renewable energy sources for countries with substantial natural wind resources and the necessary technological infrastructure and support policies. In well-developed, relatively small, homogeneous countries such as Denmark, Sweden, Ireland, and Portugal, wind energy could meet up to half of total electrical needs. For larger countries, such as China, India, and the United States, wind energy could reach 20–30% of total generating capacity, but probably not much more is realistically feasible given their diverse geographic, demographic, and economic conditions. For all countries, capacity-resource generation based on fossil and nuclear fuels will be necessary for the foreseeable future to meet increasing baseload and peak-load electricity demand.

Building Sustainable Energy Futures

The Merriam-Webster definition of the term sustainable refers to the ability to last or continue for a long time. The term's more commonplace usage in policy and science harks back to its introduction in the United Nations'-commissioned

Brundtland Report on Environment and Development, which argued for a new approach toward economic growth and development (Brundtland 1987). Sustainable development recognizes the critical interactions between natural and social systems that links human well-being with an environmentally conscious approach to economic development, and that, above all, focuses on meeting the needs of current and future generations.

The ubiquitous use of the term sustainability today suggests not only a deep resonance with the public, but more importantly the education of a generation since its introduction. Moreover, the now generally accepted interdisciplinary academic field of sustainability science provides the framework for a more rigorous and problem-driven, science-based study of general ideas encapsulated within the term (Kates et al. 2001; Clark and Dickson 2003; Clark 2007).

The global recognition of the role of conventional fossil fuels and greenhouse gas emissions in human-induced climate change has prompted a strong push for the use of alternative, more benign and renewable, sources of energy, of which wind energy has become a major resource (Aber et al. 2015). Energy demand will continue to increase as such countries as China, India, and Brazil fuel their social and economic development goals. Even with calls for a sustainable path to development, the reality for these and other developing nations is to meet the current needs of their citizens. Even so, alternatives to the current energy system do exist and are now being pursued to varying extents by countries across the globe (Fig. P3). Countries also recognize the distinct local advantages of investing in renewables for regional economic growth (Lewis and Wiser 2007). Not only does it help improve energy access and security, in the long run it may also help reduce the dependency on fossil fuels.

A multi-pronged strategy, the so-called **radical middle**, could help secure a country's energy future (Tinker 2013). Energy productivity may be enhanced through promoting conservation, securing efficiencies in manufacturing, transportation, and household use. At the same time, however, a pronounced shift to renewable sources such as wind energy, where appropriate, should also be emphasized in order to provide for a balanced and sustainable energy portfolio that exploits the strengths of each type of energy resource.

At present, the wind-energy sector requires long-term governmental commitments through the implementation of appropriate policies, the availability of financing, and the creation of markets, among other factors (Dincer 2011). In the near term, strategic investments in research and development, infrastructure, and implementation also would be needed. The rapid expansion of wind energy in many countries provides evidence of such policies contributing to both regional development strategies as well as overarching goals for renewable energy production (Fig. P4).

Figure P3 Wind turbines on Ascension Island, an isolated space-command and air-force site in the South Atlantic midway between Africa and South America. Lance Chung/Wikimedia Commons/Public Domain.

Figure P4 Recharging a Renault EV minivan for local delivery service in Aarhus, Denmark. A scene that is increasingly common around the world.

References

Aber, J.S., Aber, S.W. and Pavri, F. 2015. Windscapes: A global perspective on wind power. Multi-Science Publishing, UK, p. 245.

Brundtland, G.H. 1987. Our common future. World commission on environment and development. United Nations, Department of Economic and Social Affairs. Sustainable Development. https://sustainabledevelopment.un.org/content/documents/5987our-common-future.pdf

Clark, W. 2007. Sustainability science: A room of its own. *Proceedings of the National Academy of Science* 104, p. 1737–1738.

Clark, W. and Dickson, N. 2003. Sustainability science: The emerging research program. *Proceedings of the National Academy of Science* 100(14), p. 8059–8061.

Dincer, F. 2011. The analysis on wind energy electricity generation status, potential and policies in the world. *Renewable and Sustainable Energy Reviews* 15, p. 5135–5142.

GWEC. (2024). GWEC global wind report 2024. Global Wind Energy Council. https://www.gwec.net/reports/globalwindreport

Kates, R., Clark, W., Corell, R., Hall, J.M., Jaeger, C., Lowe, I., McCarthy, J., Schellnhuber, H., Bolin, B., Dickson, N., Faucheux, S., Gallopin, G., Grubler, A., Huntley, B., Jager, J., Jodha, N., Kasperson, R., Mabogunje, A., Matson, P. and Mooney, H. 2001. Sustainability science. *Science* 292(5517), p. 641–642.

Lewis, J.I. and Wiser, R.H. 2007. Fostering a renewable energy technology industry: An international comparison of wind industry policy support mechanisms. *Energy Policy* 35(3), p. 1844–1857.

Tinker, S. 2013. Energy 360: Leaving our corners for the radical middle: New Zealand sets an example. *Earth* 58(9), p. 8–9.

Acknowledgments

Many individuals contributed to the preparation and completion of this endeavor. Without their help this book hardly would have been possible. In particular, we wish to thank Fred Anderson, Gayla Corley, J.T. Elder, Tim Esser, Maria Górska-Zabielska, Preben Jensen, Susan Kelly, Dave Leiker, Holger Lykke-Andersen, Irene Marzolff, Chris Pettit, Jan Piotrowski, Jill Schramm, Jim Schubert, C. Shepherd, Megan Sprague, Cheryl Unruh, and Elmer Værge.

Financial support was provided by institutional grants from Emporia State University, Kansas; San José State University, California; and the University of Southern Maine in the United States.

Unless otherwise noted, all photographs, images, and diagrams were created by the authors.

1

History of Wind Energy

The fast growing world-wide wind industry is basically using the concept developed in Denmark from 1975 to 1979 (Maegaard 2009, p. 51).

1.1 Introduction

People have utilized the wind to power various types of devices and vessels since the dawn of civilization. Primary applications prior to the twentieth century were pumping water, grinding grain, and sailing ships. In all cases, some type of airfoil is utilized to capture the kinetic energy of the wind and convert that energy into some kind of mechanical action that produces work useful for human activities. Various rigid or soft wings, blades, sails, chutes, vanes, fins, and tails serve to generate lift and to provide stability.

Sailing was and continues to be the most widespread use of wind power in nearly all countries and cultures around the globe. The original scale for wind speed was devised for this purpose in 1805 by Sir Francis Beaufort of the British Royal Navy (Table 1.1). Although modified for other units, this wind scale remains the global standard.

Windmills of many types and sizes are ubiquitous around the world primarily for pumping water, grinding grain, sawing wood, and running diverse machinery. Windmills may have been used in Persia as early as the seventh century and likely by the tenth century AD (Beedell 1975; Musgrove 2010). These were somewhat similar to waterwheels turned on side, so the wheel rotates horizontally on a vertical axle. The Chinese likewise have used horizontal windmills since at least the thirteenth century (Fig. 1.1).

Wind Energy Landscape: Principles, Techniques, History, First Edition. James S. Aber, Susan E. W. Aber, and Firooza Pavri.

Published 2026 by John Wiley & Sons, Inc.

Table 1.1 Beaufort wind scale for categories 0–10 in relation to wind energy.

Rank	Wind*				Beaufort	Surface effects	Wind energy
	knots	m/s	km/h	mph			
0	<1	0.5	1.8	1.2	Calm	Smoke rises straight up, sea and lake mirror-like surface	Turbines shut off
1	1–3	1.5	5.4	3.5	Light air	Smoke drifts, wind cannot be felt, smooth water surface	Turbines shut off
2	4–6	3.0	10.8	6.9	Light breeze	Wind felt on face, leaves rustle, ripples on water surface	Minimal
3	7–10	5.1	18.0	11.5	Gentle breeze	Leaves and twigs flutter, small flags extended, small waves	Moderate
4	11–16	8.2	29.1	18.4	Moderate breeze	Wind raises dust, branches move, small waves, numerous whitecaps	Good
5	17–21	10.7	38.2	24.2	Fresh breeze	Small trees sway, many whitecaps, some spray	Good
6	22–27	13.9	49.1	31.1	Strong breeze	Branches move, lines whistle, whitecaps everywhere	Excellent
7	28–33	17.0	60.1	38.0	Near Gale	Whole trees moving, resistance felt walking against wind	Excellent
8	34–40	20.6	72.8	46.0	Gale	Large trees move, walking difficult, large waves, howling sound	Excellent
9	41–47	24.2	85.5	54.1	Strong gale	Slight structural damage occurs, slate blows off roofs	Dangerous
10	48–55	28.3	100.1	63.3	Storm	Trees broken or uprooted, considerable structural damage	Turbines shut off

Range of wind speed (*) for each category given in knots (nautical miles per hour); equivalent upper limit for each category also given in meters per second (m/s), kilometers per hour (km/h), and miles per hour (mph). Wind energy is indicated for operation of large, modern wind turbines. Aber et al. (2015).

Figure 1.1 Sketch of traditional Chinese windmill for pumping water. Carl von Canstein / Wikimedia Commons / CC BY-SA 3.0.

1.2 Traditional Windmills–European and American

Traditional windmills first came to Europe around 1100 and spread from France to other regions (Thorndahl 2005). The earliest written reports date from 1180 in France and 1185 in England, and the oldest illustration comes from 1270 (Beedell 1975). In all cases, these were so-called post mills (Fig. 1.2). The sails rotate on a wheel in a near-vertical position on a near-horizontal axle, and the whole body may be turned on a central post to face the wind in any direction. This arrangement is clearly quite different from Persian or Chinese windmills, which suggests the European windmill was an independent innovation.

From this beginning, European windmills developed into much larger and more complex devices including tower, smock, and polder mills. During the next several centuries tens of thousands of windmills were constructed across Europe, and the technology was exported to America and elsewhere. The era of windmills lasted until the early twentieth century, some eight centuries, which is a remarkable achievement for any human technology (Fig. 1.3).

All traditional European windmills operate on the same basic principles. The sails of the wheel face into the wind and act much like ship sails or airplane wings. The common sail consists of cloth stretched over a wooden framework that is inclined at an angle (pitch) to the plane of the wheel. The pitch of the sail causes it to drive forward and, thus, turn the wheel. In early windmills, the sails were set at a fixed pitch of ~20° (Beedell 1975). However, the outer tip of the sail travels much faster than the inner portion. It was discovered that pitch could be twisted along the sail to take advantage of this. The so-called weathered sails have pitch ~20° in their inner portions and flatten to pitch of only ~5° toward their tips.

Figure 1.2 Egeby Stubmølle (post mill) was built in 1787. The mill is rotated manually with the long tail pole (to right). It is no longer in service, but has been restored and functional since 1999. Island of Bornholm, Denmark.

Figure 1.3 Klostermøllen near Vestervig in northwestern Jutland, Denmark. A typical Dutch-style windmill; it was built in 1860 and continued in service until 1960. The mill was restored to its original state in the late 1980s. Further restorations were necessary after storm and lightning damages.

The rate of rotation for practical work is typically 12–15 revolutions per minute (rpm) or one full rotation every four to five seconds at wind speeds of 4–10 m/s (see Table 1.1). More rapid rotation could lead to structural damage to the windmill. Hence, some means are necessary to control excessive speed in strong wind (>15 m/s). This may be accomplished in several ways. The simplest way with early post mills was to rotate the sails parallel to the wind. A mechanical brake was soon added that allowed the miller to slow and stop the wheel with one sail in the six o'clock (downward) position. The cloth cover then could be reefed (folded back) to reduce its surface area (Fig. 1.4). This had to be done for all sails to accommodate the wind speed, a laborious process that might be repeated several times during the day to adjust for changing wind conditions.

Beginning in the late eighteenth century, spring sails and patent sails were developed with shutters or flaps that were adjusted automatically using springs or weights to govern the rate of rotation (Beedell 1975). The so-called flap-shutter blades allowed the miller to adjust easily for variable wind speed and to turn the mill on and off at will. Another important innovation was the fantail. Positioned on the back side, the fantail is mounted vertically at right angle to the main wheel.

Figure 1.4 Close-up view of windmill wheel, Greek island of Mykonos in the Aegean Sea. The canvas sails are mostly reefed to reduce rotation speed. The earliest of these windmills date from the 1500s. Note the elaborate construction with braces and stays to reinforce the wheel. Similar windmills are known from other countries, such as Portugal (Autocar 1961). This technique was repeated on Danish wind turbines in the mid-twentieth century. Photo by R.K. Aber.

When the wind shifts to one side, the fantail begins to spin, and it drives a mechanism that turns the mill cap automatically into the wind again.

Fantails and flap-shutter blades represent the ultimate development of traditional European windmills by the turn of the twentieth century (Fig. 1.5). Typical mills of this type utilize only about 6% of potential wind energy (Thorndahl 2005). Such mills had an average power output of ~10 kilowatts (kW) with a maximum up to 40 kW (Musgrove 2010). At 10 kW, the mill was equivalent to the muscle power of 150–200 people.

Traditional European-style windmills also were built across the United States and Canada. A general impression is that such windmills were common only along the eastern seaboard, but they were actually quite numerous from coast to coast (Fig. 1.6). While European-style windmills were common, American windmill innovation took a completely different direction beginning in the mid-1800s. The prototype multi-bladed windmill was invented by Daniel Halladay in 1854 at Ellington, Connecticut (Baker 1985). Rapid development of diverse designs by many companies followed immediately.

Figure 1.5 Aarsdale Mølle, a smock, Dutch-style windmill on the island of Bornholm, Denmark. (Left) Flap shutters in closed position and blades turning in counterclockwise direction, (Right) flap shutters in open position and blades not turning. Note variation in pitch along length of blades and the fantail on back side of mill. This mill was built in 1877, and the original sail blades were replaced with flap-shutter blades in 1919–1921. The mill still operates today, although not on a regular basis.

Figure 1.6 Old Dutch Mill in Wamego, Kansas. The stone-tower mill was built in 1879 by a Dutch immigrant, J.B. Schonhoff, and used for grinding wheat and corn. The mill was taken down and rebuilt in Wamego in 1925 as a historical monument. The sail frames are non-functional decorative structures.

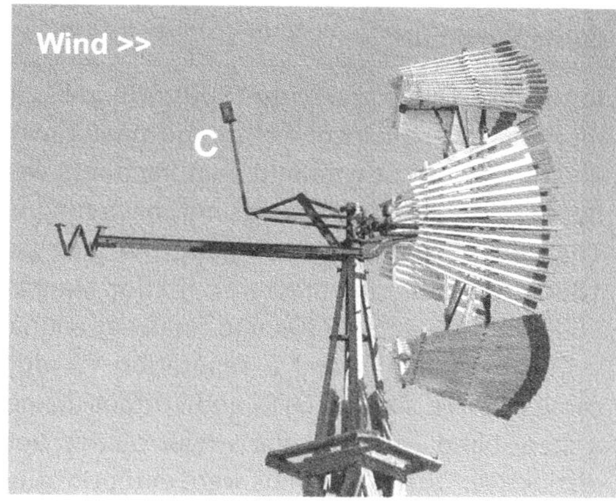

Figure 1.7 Side view of restored Raymond vaneless windmill at Lincoln County Museum, Kansas. The W counterbalance arm points into the wind, and the blade sections are tilted in the downwind direction. The counterweight (C) controls the tilt of the blades; shown here in a position for strong wind. This model was among the most common vaneless windmills across the U.S. Midwest and Great Plains in the late 1800s and early 1900s (Baker 1985).

The early windmills were mostly of the sectional-wheel type in which sets of closely spaced blades fold backward as wind speed increases, thereby controlling the rate of rotation. Some type of spring or counterweight governs the tilt of the blades. Some designs employed a large tail vane to point the bladed wheel into the wind; others were vaneless (Fig. 1.7). The main alternative to sectional wheels is the solid wheel in which all the blades are held rigidly and do not fold. Rate of wheel rotation is governed in various ways with a side vane, offset vane, or off-centered wheel.

Many variations of both sectional- and solid-wheel windmills were manufactured during the late 1800s and early 1900s. Early versions were made largely of wood with wheel diameters typically in the range 2–9 m and giant windmills up to 18 m in diameter (Baker 1985). The superiority of steel began with systematic experiments in 1882–1883 by Thomas Perry at the U.S. Wind Engine and Pump Company in Batavia, Illinois. Based on thousands of tests, he devised the now-familiar wind wheel (wind rose) with curved steel blades set at a specific angle within a rim that offered minimal resistance to the wind (Fig. 1.8). This wheel proved far more efficient than existing wooden wheels. Perry joined forces with LaVerne Noyes to form the Aermotor Company in 1888, which in relatively short time became the largest and best-known maker of American windmills.

In contrast to European windmills, American windmills have many blades in their wheels, at least a dozen to >100 blades in some models. According to Musgrove (2010), *the number of blades has little effect on the power output, for the power in the wind flowing through the rotor can be efficiently intercepted using any number of blades* (p. 53–54). However, the number of blades does influence torque and minimum wind speed needed to start the wheel turning. More blades result

Currie Windmills

In the late 1800s and early 1900s, windmills in Kansas pumped groundwater for railroad steam locomotives and small towns as well as thousands of farms and ranches. Most windmill manufacturers were located in states to the east, but as many as 50 companies may have built windmills in Kansas (KHS 2014). The Currie windmill was among the best known and long lived from the 1880s to 1950s, manufactured first at Manhattan, then Topeka from around 1900 until the late 1940s, and finally at Salina.

Currie windmills had a reputation for durability. The wind-rose wheels came in three sizes of 6-, 8-, and 10-foot diameters. Bearings were hard wood, a steel band encircled the wheel blades, and the tail vane was large and distinctive. Currie windmills were exceptionally low in cost and became known as the *poor man's windmill* (KHS 2014). They were quite common in Kansas as well as many other states. Currie windmills are preserved in several museums today, and the tradition continues with modern mini-Currie windmills made for garden displays and weathervanes.

Historic farmstead of D.R. Beckstrom, who moved to Greeley County in westernmost Kansas in 1892 (C. Shepherd, pers. com. 2023). The Currie windmill has a large wheel with 30 blades. Notice the wooden water barrels at the bottom. The mill presumably dates from the 1890s, as it appears to show Manhattan as the manufacturing city. Undated image obtained from the Horace Greeley Museum in Tribune.

in a higher starting torque at low wind speed, which is particularly useful for operating reciprocating water pumps. Aermotor and similar geared windmills could pump water at wind speeds as low as 2 m/s (Baker 1985).

The success of the American windmill was exported and copied widely in other countries, particularly Argentina, Australia, Brazil, Canada, China, India,

Figure 1.8 Baker Monitor W Series, back-geared, self-oiling windmill restored at Lakin, Kansas. The wheel has 15 fixed, curved steel blades. The distinctive vertical vane was introduced around 1940 and continued to be made into the 1960s. Note the bladed wheel is slightly off centered; as wind speed increases the wheel tends to turn toward the vane, thus regulating the rate of wheel rotation. As wind speed decreases, a governor spring pulls the wheel to face the wind again (Baker 1985).

Mexico, New Zealand, South Africa, Uruguay, and Venezuela, to name a few. Large windmills of the Halladay type were designed, built, and installed in Denmark by N.J. Poulsen in the 1870–1880s (Christensen 2009). By 1930, as many as six million American windmills had been built (Musgrove 2010). However, the Great Depression of the 1930s was a blow that many windmill companies could not survive, and American windmills reached a low ebb in the mid-twentieth century.

This situation turned around with sharp increases in energy costs beginning in the 1970s and since. Water-pumping windmills once again became common for both new equipment and maintaining older models in service. Also restoring antique windmills became popular (Fig. 1.9). The long-term success of the American windmill is due to several factors, most importantly American windmills are self-regulated and run autonomously with minimal human attention (Baker 1985).

1.3 Generating Electricity–Late 1800s to Mid-1900s

Electricity is ubiquitous in the modern world; so much so, that we take it for granted nearly everywhere. We simply "plug in" to the wall socket, put batteries into our portable devices, or connect the car to a fast-charging station. Yet widespread availability and use of electricity are hardly more than a century old.

Figure 1.9 Jasper Windmills, an open-air exhibit in southwestern Minnesota, restores and displays antique windmills from the United States and several other countries including some rare and unusual types.

The concept of utilizing the wind to generate electricity goes back at least to the 1860s. For example, several successful efforts were made by Prof. James Blyth of Anderson's College (now Strathclyde University), Scotland, in 1887–1914, and in the United States by Charles Brush of Cleveland, Ohio, in 1888–1908 (Price 2005; Musgrove 2010).

The most important of these early efforts with long-term significance took place in southwestern Denmark under the leadership of Poul la Cour (1846–1908). A scientist, teacher, and inventor, he obtained governmental support to build an electricity-generating windmill at Askov in 1891. A second, larger windmill was constructed in 1897, on the basis of extensive experiments in a wind tunnel he designed. La Cour formulated many of the basic tenets of modern wind power, and his protégé, Johannes Juuls, innovated the basic concepts of modern wind turbines half a century later.

The period 1900–1920 is known as the first "golden age" of wind energy in Denmark (Christensen and Thorndahl 2012). By the end of World War I, approximately one-quarter of rural Danish power stations relied on wind turbines of the la Cour type. When la Cour's 1897 Askov windmill was destroyed by fire in 1929, it was replaced with a Lykkegaard model that had a decidedly modern look—four metal-covered blades mounted on a steel-truss tower (Musgrove 2010).

World War I stimulated rapid aeronautical developments for wings and propellers. These concepts were applied soon to Danish wind turbines in the form of the Agricco windmill, which proved to be considerably more efficient than

Poul la Cour

Danish developments were in progress under the leadership of Poul la Cour (1846–1908). A meteorologist, teacher, and prolific inventor, known as the Danish Edison, he obtained governmental funding to build an electricity-generating windmill at Askov in southwestern Denmark in 1891. Batteries were unreliable and expensive at the time, so his idea was to use the electricity to electrolyze water and produce hydrogen and oxygen gases, which could be stored and later burned for lighting (Nissen 2009; Christensen and Thorndahl 2012).

The first windmill, built by N.J. Poulsen, was quite traditional with four canvas-covered sails and two fantails. These sails were soon changed to adjustable shuttered or flap-blades; the flaps opened automatically if the wind was too strong. In 1897, la Cour teamed with the architect P.V. Jensen-Klint and the millwright Christian Sørensen to design and build a larger windmill with further funding from parliament (Christensen 2017). Sørensen had developed a "conical wind rotor" which was installed with six blades.

Windmills built by Poul la Cour for generating electricity at Askov, southwestern Denmark. First mill (right) was erected in 1891 and had a diameter of 11.6 m with four flap blades. The second mill (left) was built in 1897 and was 22.8 m in diameter. The picture dates from that year and shows the six-bladed conical wind rotor of Christian Sørensen.

(Continued)

Poul la Cour (Continued)

However, this setup gave so many problems that la Cour switched after only eight months to a more-conventional four-blade design (Thorndahl 2005). This led to a long and bitter debate between Sørensen and la Cour (Quistgaard 2009), but investigation showed the four-blade design was more effective than the conical wind rotor.

La Cour invented many mechanical and electrical devices for improved capture and storage of wind energy during the last decade of the nineteenth century. He tested the aerodynamics of windmill blades and model windmills in wind tunnels and confirmed the long-held belief that the blade-swept (rotor) area was a key factor for the mill's working capacity (Nissen 2007). Based on his experiments, la Cour formulated the classic relationship:

$$\text{Work} = \text{constant} \bullet \text{blade} - \text{swept area} \bullet \text{wind speed}^3$$

The cube of wind speed means that a slight increase in wind speed yields a large increase in electricity generation. He introduced the American blade type with flaps. Traditional Dutch windmills utilize only about 6–8% of potential wind energy, but la Cour's mill utilized 23% (Thorndahl 2005). He furthermore discovered that fewer blades (4) gave greater power output than more blades (6, 8, 16). On the basis of wind-tunnel testing, la Cour developed principles for his *ideale mølle* (ideal windmill). La Cour had a social vision that wind-generated electricity would help modernize rural Danish life in the early twentieth century.

In order to promote this goal, he offered a prize in 1901 for the ideal mill. Two prize winners were millwrights Ole Jensen and Niels Hansen. The latter began producing windmills under the name Lykkegaard, and this firm continued to manufacture windmills in the la Cour tradition until 1957 (Christensen and Thorndahl 2012). Lykkegaard mills were also exported to France and Venezuela. During WWII (1940–1943), some 60 large flap-blade windmills were installed for generating electricity throughout Denmark (Christensen 2017). The building that supported la Cour's 1897 windmill survives as the Poul la Cour Museum.

traditional windmills or Lykkegaard turbines (Table 1.2). However, ready availability of imported oil and coal following World War I lessened the demand for wind power; production of the Agricco ended in Denmark in 1926, and manufacturing rights were sold to a Dutch company that continued production a few more years (Christensen 2008).

Table 1.2 Efficiency of Danish wind turbines for harnessing potential wind energy, early and mid-1900s.

Turbine type	Date	Blades and rotor (m)	Capacity (kW)	Turning method	Air brake method	Efficiency (%)
La Cour-type Lykkegård	1914–1945	4 blades and 7–18	5–30 DC	Fantail	Flaps in blades	20–25
Jensen & Vinding Agricco	1918–1925	4–6 blades and 5–12.5	5–40 DC	Fantail	Blade pitch control	40–45
F.L.S. Aeromotor	1940–1945	2 or 3 blades and 17.5–24	50–70 DC	Fantail	Spoilers on blade backs	30–40
Juul's Vester Egesborg	1950	2 blades and 7.65	15 AC	Electric motor	Flaps at tips of blades	40–50
Juul's Bogø	1952	3 blades and 13	45 AC	Electric motor	Flaps at tips of blades	50–55
Juul's Gedser	1957	3 blades and 25	200 AC	Electric motor	Flaps at tips of blades	40–50

Thorndahl (2005, Tables 1 and 2, p. 14–15).

Sporadic efforts were made elsewhere early in the twentieth century to harness wind energy for generating electricity. One of these took place at San Gorgonio Pass, California, a site that now hosts a large wind-energy complex. Dew Oliver, a colorful Texas inventor, built what he called a "blunderbuss" at the pass in 1926 (Baker 1985). He claimed it worked, but his scheme to build hundreds of similar turbines and power all of southern California eventually failed, and he was convicted of fraud. Another large and more successful wind turbine of this vintage was the Russian 100 kW generator at Balaclava on the Crimean Peninsula in the northern Black Sea. It was constructed in 1931 and operated for a decade until it was damaged in World War II (Musgrove 2010).

Marcellus and Joseph Jacobs began experiments to make a small, efficient wind generator in the 1920s on their Montana ranch. They soon arrived at a three-bladed configuration with a tail vane; rotational speed was controlled by variable blade pitch that was adjusted by a centrifugal governor (Musgrove 2010). The primary role was to charge batteries for operating light bulbs, radios, and small appliances. The company moved to Minneapolis in 1931, and some 20,000 small wind generators were sold before production ceased in the late 1950s. However, production restarted in 1980 and continues into the twenty-first century (Fig. 1.10; Jacobs 2019).

Agricco Windmill

Danish windmills had a renaissance and new development of blade profiles as a result of World War I, namely the propeller mill patented by the engineers Johannes Jensen and Poul Vinding. The *Agricco-mølle* was presented in 1919. It represented a new generation of windmills that featured blades based on airplane wings. Blades had steel frames with wooden ribs, and the turbine was self-directed into the wind. Rotor speed was adjusted by turning the entire blade, in other words, pitch regulation. Rotor diameter varied from 8.5 to 16.5 m, and blade number was four, five, or six.

The aerodynamic design of windmill blades in Denmark was well ahead of other countries around 1920 in terms of both techniques and scientific understanding. Government testing in 1923 showed that Agricco mills were more efficient at high wind than la Cour's ideal mill and at low wind were more effective than wind-rose rotors. Agricco windmills utilized 43% of potential wind energy (Thorndahl 2005).

An asynchronous generator was perfect for windmill operation to produce AC current directly into the grid system, although only a few Agricco mills were sold for electricity production. The timing of Agricco was unfortunate, however, as the market for new mills was on the way down after World War I, and the factory was liquidated in 1925 (Christensen 2017). Agricco windmills proved robust, and some continued in service until the 1960s, although only a few skeletons remain today.

Agricco windmill on the island of Bornholm, Denmark. Five-bladed model with twin fantails erected on the base of an older Dutch-style mill in Olsker. This mill was in service from 1927 until the late 1960s (Christensen 2016). Seen here as it appeared in 1979.

Figure 1.10 Jacobs wind turbines in Hawaii circa 1989. Paul Gipe / Wikimedia Commons / CC BY-SA 4.0.

In the 1930s, Palmer C. Putnam, a New England engineer and geologist, became interested in industrial-scale wind power for generating electricity. He convinced the S. Morgan Smith Company to finance the construction of a large wind turbine at Grandpa's Knob in the Green Mountains of Vermont (Baker 1985; Musgrove 2010). The turbine had twin blades with a rotor diameter of 175 feet (53 m) standing on a steel-truss tower, by far the largest wind turbine built up to that time with a maximum output of 1.25 megawatts (MW). The Smith–Putnam wind turbine began operation in 1941, but it suffered numerous mechanical failures and was eventually abandoned shortly after World War II.

The war and its immediate aftermath marked a time of severe restrictions for materials, fuel, and labor in many countries, which curtailed further wind-energy developments during the 1940s. The war-time need for wind power stimulated the F.L. Smidth & Co.'s Aeromotor development in Denmark. Designed by Helge Claudi Westh (1904–1992), these two- and three-bladed turbines were mounted on reinforced concrete towers (Thorndahl and Christensen 2009). The largest turbines had rotors 24 m in diameter with 70 kW capacity. These turbines proved quite significant for subsequent work by Johannes Juul in the 1950s.

Johannes Juul (1887–1967) took up the cause for further development of wind energy. Juul had been a student of la Cour's in the first class for wind electricians at Askov in 1904 (Christensen and Thorndahl 2012). He conducted wind-tunnel experiments and built a series of prototype turbines that culminated with the Gedser turbine in southeastern Denmark. The Gedser wind turbine was put into continual, unattended operation in 1958, and it generated grid power for a decade. Juul's innovations proved to be pivotal for modern turbine design (Thorndahl 2009), and without question the Gedser turbine had great influence for continued development of the wind industry in Denmark and elsewhere (Musgrove 2010).

Johannes Juul and the Gedser Wind Turbine

A direct line extends from Poul la Cour's windmills at Askov to modern Danish electricity generating wind turbines via Johannes Juul (1887–1969), who was a pupil in the first course for rural electricians in 1904 (Thorndahl 2005). From 1905 to 1925, Juul held various electrical positions in Denmark and Germany. In 1926, he accepted a position with SEAS (Sydøstsjællands Elektricitets Aktieselskab), where he had the freedom to conduct independent research.

In 1947, at age 60, Juul began the work that would make him world famous in the wind-energy community. He built a wind tunnel in 1948 to test blade profiles. At least 25 different blade types were tested, which showed that quite small deviations had great significance. Air brakes and spoilers could be integrated into blades. Juul reached conclusions in 1949 about the ideal blade and mill. From numerous sketches, he clearly favored a three-bladed rotor with brake flaps in the blade tips.

The first research wind turbine was erected at Vester Egesborg in 1949, and a second mill was built in 1952 at Bogø, both in southeastern Denmark. Blade regulation had proven to be the weak point with contemporary windmills in England and the United States. Juul's solution with brake flaps in the blade tips was simple, less expensive, and more stable than pitch regulation for the whole blade. Juul had solved the problem of regulating rotation speed, and in 1952 he got a patent for this technique.

In 1954–1955 plans were developed for a large 200 kW research wind turbine, and Parliament provided money from the Marshall Plan for wind-energy research (Thorndahl 2005). Gedser, at the southern tip on the island of Falster in southeastern Denmark, was chosen as the location, because personnel at SEAS had the most experience with the previous experimental windmills.

Much foreign interest was shown for *Gedsermøllen* when it was completed in 1957. The main aspects included a 24-m-diameter rotor with three blades and two braking systems, namely blade-tip brake flaps and a mechanical brake. The 200 kW asynchronous generator was designed to complement blade size and wind conditions in the vicinity. The tower was 25 m tall made of concrete with three buttress ribs on the sides. All aspects of mill design and construction were compromises in order to minimize cost.

The windmill opened in 1957, the same year that Juul retired. After some minor adjustments and measurements the mill was fully operational in spring 1958. The Gedser turbine ran continuously until 1967 and produced 2,240,000 kWh of electricity (Thorndahl 2005). Juul concluded in 1962 that wind energy could:

Johannes Juul and the Gedser Wind Turbine (Continued)

- Save cost of generating electricity over foreign fuels.
- Achieve reserve energy for steam-power plants when weather is windy and cold.
- Create a work market for building of wind-power facilities.
- Achieve economic distribution by Danish wind power working together with Norwegian and Swedish hydropower.
- Create a market for export of Danish wind-power facilities to the rest of Europe.

It took two more decades to prove that Juul was correct. However, he died in 1969 without realizing the full potential of his vision. State support for wind energy research ended in 1962. Interest in wind energy was minimal in Denmark at this time. Fossil fuels were cheap, and nuclear energy was just around the corner. In 1965 Jylland and Fyn, and earlier Sjælland, were connected via undersea cables to cheap and reliable Swedish hydropower.

Gedser turbine nacelle and blades restored and displayed at the Danish Energy Museum. Blades have steel beams with wooden cross ribs, and the blade surface is covered with aluminum sheets. Turbine blades have cross braces and struts for extra support; these braces were oversized to protect the blades from storm damage. Note the blade-tip flap brakes.

(Continued)

Johannes Juul and the Gedser Wind Turbine (Continued)

All this changed with the oil crisis in 1973. The Gedser turbine was the only large mill in the world that had operated without significant interruption for 10 years. DEFU (Dansk Elværkers Forenings Udredningsudvalg) and ERDA (U.S. Energy Research and Development Administration) agreed to restore the Gedser mill, which was renovated in 1977 and ran again until 1979. Thus, the Gedser turbine spanned more than 20 years of operation and measurements.

The Gedser wind turbine represents the **Danish Concept**, which includes a three-bladed rotor mounted upwind from the tower, dual aerodynamic and mechanical braking systems, and asynchronous AC generator connected directly to the electricity grid. In the 1970s, Gedser was the starting point for further development of both large and small Danish wind turbines.

Key criteria behind Juul's construction at Gedser were simplicity, safety, and low cost. Gedser was a success in comparison to contemporary ambitious wind turbines in other countries. Those elegant turbines suffered many technical problems, but Gedser was not elegant and it did work. Juul had carried out practical experiments and had upscaled smaller mills into the large Gedser turbine. Gedser represents today one of the most important technological Danish innovations since WWII and the basis for the modern worldwide wind industry.

Like Denmark, France, Germany and the United Kingdom experimented with AC-generating, grid-connected wind turbines during the 1950s and 1960s (Musgrove 2010). Several large wind turbines were designed and tested at inland and coastal sites as well as small islands, such as the Orkneys and Isle of Man. A noteworthy series of turbines was designed by Lucien Romani in France (Cavey 2024). He founded the Bureau des études scientifiques et techniques (BEST) in 1946 and produced a series of increasingly large wind turbines known as BEST–Romani aerogenerators.

Among the most remarkable turbines of this era were those developed by Ulrich Hütter at the Technischen Hochschule Stuttgart (now University of Stuttgart), Germany. Hütter was an aircraft designer who began field-testing wind turbines based on propeller designs in 1940. Continued experiments resulted in a 34-m-diameter, two-bladed turbine in 1957 (Fig. 1.11). He was the first to employ fiberglass to construct self-supporting blades that were extremely long and thin. Hütter's work had decisive influence on subsequent Danish developments at Tvind.

At this time, large turbines were quite few in number, located in relatively isolated spots, and generally did not attract much attention from the public in terms of their

Figure 1.11 Allgaier–Hütter StGW-34 model, two-bladed, 100 kW wind turbine located at Stötten, ~80 km east of Stuttgart, Germany (1958). With a rotor diameter of 34 m, it was the first turbine with fiberglass blades. Courtesy of H.H. Dörner.

appearance or potential environmental impacts. In fact, the environmental movement as we know it today hardly existed. Most international treaties and national laws regarding air, water, wetlands, wildlife, pollution, climate, and other environmental issues were still in the future. Thus, the emphasis on developing wind energy was focused almost entirely on practical and economic questions of electrical and aerodynamic engineering.

By the 1960s, however, vast oil reserves of the Middle East were flooding the world with low-cost fuel, nuclear energy was growing, and the perceived need for wind energy waned (Musgrove 2010). The Gedser wind turbine was shut down in 1967. The Allgaier–Hütter, BEST–Romani, and other large wind turbines suffered similar fates, as the general public and policy makers firmly believed that cheap oil and nuclear energy would be unlimited and have no environmental consequences.

These assumptions soon would change dramatically. When the United Kingdom withdrew from the Persian Gulf region in 1971, the politics of petroleum changed, and the first great oil crisis was triggered barely two years later. A second oil crisis in 1979 and nuclear disasters at Three Mile Island, Pennsylvania, in 1979 and Chernobyl, Soviet Union (now Ukraine), in 1986 further drove public opinion and policy decisions regarding energy resources. Yet, it would take much of the late twentieth century for the modern wind industry to emerge as a leading source for renewable, non-polluting energy.

1.4 Generating Electricity—Late 1900s

Interest in and development of wind energy had reached a low ebb by the end of the 1960s. The Yom Kippur War in 1973 triggered OPEC to reduce oil production, which led to shortages and rapid price increases, and high oil prices continued through the 1970s. Many countries recognized the need to develop renewable, domestic, safe, and non-polluting forms of energy. Research was conducted on many potential energy sources, namely solar, geothermal, wave and tidal, and wind. In the 1970s, however, wind was considered an unlikely prospect for generating substantial electrical power compared with other renewable energy sources.

1.4.1 Developments in the 1970s

Several relatively small wind-energy programs were started nonetheless in Denmark, Germany, Netherlands, Sweden, United Kingdom, and United States (Musgrove 2010). Among the most important wind-energy developments during the 1970s and 1980s were those in Denmark and the United States. The Danish government and electric utilities wanted to build nuclear power stations as quickly as possible. However, many grass-roots organizations sprang up in opposition to nuclear energy and in support of renewable energy sources, namely solar and wind.

Lars Albertsen proposed a new Danish word, *vedvarende energi*, meaning sustainable or renewable energy (Grove-Nielsen 2023). Many Danish entrepreneurs, teachers, and inventors began to experiment with wind turbines, particularly in Jutland in the mid-1970s. Among the most important of these were small turbines built by Christian Riisager and the large turbine at Tvind.

Erik Grove-Nielsen became involved with wind turbines as a young man in 1977 to build fiberglass turbine blades, and he set up Økær Vind Energi (Grove-Nielsen 2023). In 1978 Grove-Nielsen and his wife, Tove, came to a decision that would have long-lasting consequences, namely blades that would rotate in a clockwise direction. Future Økær customers—Vestas, Bonus, Nordtank, and Enercon—followed the clockwise pattern, which set the industry standard.

Nordvestjydsk Institut for Vedvarende Energi (NIVE) was founded in Thisted in 1977 and led by the charismatic Preben Maegaard (1935–2021). The aim was to establish decentralized production of wind turbines. NIVE developed the so-called *Smedemestermølle* (Blacksmith turbine) with the idea that it could be produced in small machine shops across the country (Fig. 1.12). Økær Vind Energi supplied fiberglass blades that rotated in a clockwise direction. Like the Riisager turbine, it had twin fantails to point the main rotor into the wind. Rated output was 22 kW; later models increased to 60 and 100 kW.

Christian Riisager

Christian Riisager (1930–2008), a carpenter, was prominent among the Danish wind-turbine innovators in the mid-1970s. He already had built a small 7.5 kW turbine to provide electricity for his home garden near Herning in central Jylland. Without permission, Riisager and his wife Boe connected this turbine to the electric grid in 1975 and made their meter run backward (Thorndahl 2011). This unofficial experiment led to formal rules for individuals to sell electricity to the utility grid system.

The Riisagers teamed with Erik Nielsen in 1976, and they began manufacturing wind turbines in Herning. The blades were constructed of laminated wood by Riisager himself, and turbines were sold in sizes ranging from 10 to 45 kW (Grove-Nielsen 2023). Riisager and other windmill pioneers found parts for their machines from junk yards and auto chop-shops. Gears, brakes, crankshafts, differentials, and other parts were repurposed for windmills to generate electricity. They also found solutions from older mills, such as fantails (Christensen 2017).

Riisager turbine at the Danish Energy Museum. The rotor consists of three blades upwind from the tower and twin fantails to turn the rotor into the wind. In keeping with windmill tradition, the blades turn in the counterclockwise direction—nearly all other modern turbines rotate clockwise as viewed from upwind.

(Continued)

Christian Riisager (Continued)

Riisager turbines followed the general design of the earlier Gedser turbine (Thorndahl 2005). His turbines were, in fact, miniature versions of the giant Gedser windmill, except that his did not have brake flaps in the blade tips. Three blades rotated upwind from the tower, and twin fantails kept the rotor facing into the wind. An induction generator connected to the grid was held at constant speed, which also maintained constant rotor speed for the fixed-pitch blades (Musgrove 2010).

Riisager sold production rights in 1979 to a new company, Windmatic. He also entered into other partnerships, and Riisager-clone turbines were manufactured under such names as Erini, Oxholm, and Sonebjerg (Thorndahl 2011). By 1980 about 120 Riisager windmills were around Denmark.

Riisager sold a 55-kW turbine to the Faroe Islands in 1980, and in 1981 the Riisager family moved to Miðvágur in the Faroes and began making larger wind turbines of 90–130 kW capacity (Thorndahl 2011). These carried the Riisager name and had slimmer blades, new gears, and were tested in wind speeds up to 60 m/s. Blade materials were wood, fiberglass, and stainless steel. Finally the Riisagers made sure to patent their wind turbine.

In 1985, Riisager began to develop a 200-kW turbine, but the project encountered economic difficulty, and Riisager returned to his original avocation as a carpenter. The Riisagers did not become rich through building wind turbines, but they were instrumental in starting the Danish wind industry in 1975 and helping to develop it well into the 1980s (Grove-Nielsen 2023).

Windmatic 65 kW turbines with 14-m-diameter rotors and hub height of 18 m. Designed by Christian Riisager. In the former Los Vaqueros Windfarm, north of Livermore, California, as they appeared in 2012.

Tvind Wind Turbine

> *The importance of the building of the Tvind Turbine was immense – at a crucial time in history, where Denmark was going to decide, whether we should embark on building nuclear power stations in the country, or not* (Grove-Nielsen 2023).

In the 1970s, Denmark faced a major dilemma and engaged in political debate whether to begin development of nuclear energy. This was an intensely polarizing issue that lasted into the 1980s. The Tvind Schools had been founded as an alternative Folk High School in 1970 with governmental support as a teachers' training college. The people at Tvind, led by the charismatic teacher Amdi Petersen, were strongly opposed to nuclear power and made the decision to build a megawatt-sized wind turbine to provide power for the school complex (Maegaard 2009; Christensen and Thorndahl 2012).

A call was sent out for volunteers—both students and professionals. The basic design was modeled on Juul's Gedser wind turbine of 1957. For technical advice, they turned to Ulrich Hütter to learn about the fiberglass blades in his turbine at Stötten, Germany. They received help from several other professionals, utilized second-hand parts, and did most of the construction themselves with volunteer labor. Construction began in 1975. The base and 53-m-tall tower are made of reinforced concrete. The nacelle is made of welded steel and contains the gearbox, generator, and the hydraulic pitch-control and emergency-stop systems. Blades were lifted into position by Klose Cranes, which later became the major company for erecting turbines all over Denmark (Grove-Nielsen 2023).

The 2.0 MW Tvind wind turbine was completed in 1978, and for many years it was the largest wind turbine in active service anywhere in the world. The Tvind turbine rotor has three self-supporting fiberglass blades, 54 m in diameter, mounted downwind from the tower. The Tvind turbine *introduced the blade technology of the future. It became the industry standard and the core technology in the decisive breakthrough for Danish wind turbine producers in the following years* (Maegaard 2009, p. 48). Its blades were replaced in 1993. For this long-term achievement, the Tvind wind turbine was awarded the European Solar Prize in 2008.

In 2023, after 45 years in operation, the turbine showed serious signs of its age. Water leakage in the nacelle had caused rusting and corrosion of the generator support frame and other components. The turbine was taken offline to perform maintenance and repairs. The renovation consisted of cleaning and sandblasting with dry-ice (CO_2) pellets. The generator was lifted, and the undercarriage was replaced with new stainless-steel supports. Marine waterproofing was the final step to seal and protect the turbine (E. Værge, pers. com. 2023). It is anticipated the Tvind wind turbine will have many more decades of service.

In addition to the large turbine, the Vestjydsk Energi Kontor (Western Jylland Energy Office) was set up at Tvind in 1976 by two young engineers, Jens B. Gjerding and Søren Kahlke. The purpose was to aid and give advice to turbine

(*Continued*)

Tvind Wind Turbine (Continued)

self-builders. They designed the small 11 kW PTG turbine and its 4.5-m-long fiberglass blades. PTG followed windmill tradition and Riisager turbines by rotating counterclockwise. Furthermore, Tvind played a prominent role in Denmark's eventual decision in 1985 not to develop nuclear energy (Maegaard 2009). Without doubt, the Tvind wind turbine inspired Danes to experiment with wind power, and it stimulated numerous, independent developments by inventive self-builders.

Tvind wind turbine near Ulfborg, western Denmark, as it appeared in 2012. The concrete tower is supported by a conical foundation also of reinforced concrete. The distinctive paint scheme was applied to celebrate the turbine's 25th anniversary in 2003.

Early blades had no aerodynamic braking capability and were subject to runaway failures. Grove-Nielsen was convinced by fellow builder and customer, Henry Jørgensen, to adopt the tip-brake technique of Juul's Gedser turbine. Økær developed a new 7.5-m-long blade, under the name AeroStar, in 1980. This blade was utilized by many companies including Danregn (Bonus) and Nordtank in early California wind farms. These blades were still in service and operating 40 years later.

Figure 1.12 Økær 5-m-long fiberglass blades on a 22 kW Smedemestermølle of the Riisager type. Note the blades turn in a clockwise direction. Restored and displayed at the Renewable Energy Museum, Denmark.

Late in 1978, Økær blades with air brakes were delivered to the Herborg Vind Kraft (HVK) turbine designed by Henrik Stiesdal and built by Karl Erik Jørgensen. This turbine was similar in many respects to the Riisager style, but without fantails. The HVK employed a wind sensor and yaw motor to point the rotor into the wind, and blades rotated clockwise as viewed from upwind. The HVK turbine embodied the so-called *Danish concept* and was the basis for continued turbine development. By the end of 1979, Vestas had signed a licensing agreement for the HVK turbine and had begun commercial production.

Thus, by the end of the 1970s, Denmark had moved through an extremely rapid and innovative phase of wind-energy development. Countless experiments were undertaken by inventors and self-builders using diverse materials, varied construction techniques, and numerous turbine and blade designs. Many small companies provided a supply chain of specialized components for manufacturing and installing commercial wind turbines based on the HVK model, which emerged as the Danish standard. This standard consisted of three fiberglass blades rotating in a near-vertical plane on a near-horizontal axis upwind from the tower with an asynchronous generator supplying electricity to the grid system (Maegaard 2009). Autonomous controls and independent braking systems provided safe operation.

These efforts were aided by freely sharing technology. *No single patent or property right blocked the innovative process in this new industry* (Maegaard 2009, p. 47). Furthermore an investment subsidy of 30% for purchase of approved wind turbines was enacted in 1979, which spurred domestic sales (Madsen 2009). This subsidy gradually was reduced over time to 10% and then eventually terminated in 1989. Additional incentives included a national recommendation to local municipalities and counties to support wind energy, and access by small wind turbines to the utility grid system. This was a remarkable social achievement, in stark contrast to normal economics, that placed Denmark into a position of leadership as the wind industry grew worldwide during the 1980s.

In the United States, the first Earth Day was celebrated on April 22, 1970. This was followed by nearly 20 major statutes enacted during the decade dealing with clean air and water, endangered species, and other environmental issues. The influential physicist Amory Lovins called in 1976 for increased reliance on renewable energy resources—solar, wind, biofuels, and geothermal—for generating electricity. The premier wind research project in the United States was conducted by NASA at the Glenn Research Center in Sandusky, Ohio, beginning in 1974 (Grove-Nielsen 2023). Sponsored mainly by the Department of Energy, most of the funding went to major aerospace and electrical engineering companies (Ragheb 2023).

The result was a series of increasingly large experimental and operational turbines from Mod-0 through Mod-5B (Fig. 1.13). All employed two-bladed rotors patterned after the earlier Smith–Putnam and Ulrich Hütter designs. Some rotors were upwind and others downwind from the tower. Early models had truss towers; later ones had tubular steel towers (Fig. 1.14). In all, some 13 turbines were built and installed across the United States from Hawaii to Puerto Rico, and one went to Sweden.

The Mod turbines suffered a series of operational problems and equipment failures, and showed slight promise for eventual commercial success. The program demonstrated the difficulty of engineering successful large wind turbines by aeronautical manufacturers that had little experience in this technology. The Mod-series models, in fact, could not compete for reliability and efficiency with smaller Danish wind turbines of the time (Musgrove 2010). The Mod-series program ended in the late 1980s.

President Carter was keen to develop alternative energy resources. Federal energy legislation under his administration included three key components, all enacted in 1978—Energy Tax Act, Powerplant and Industrial Fuels Act, and Public Utilities Regulatory Policy Act (PURPA). Together these tax incentives added up to 50% in California, and PURPA guaranteed that electricity could be sold at a favorable price. U.S. Windpower, which already had built its first wind

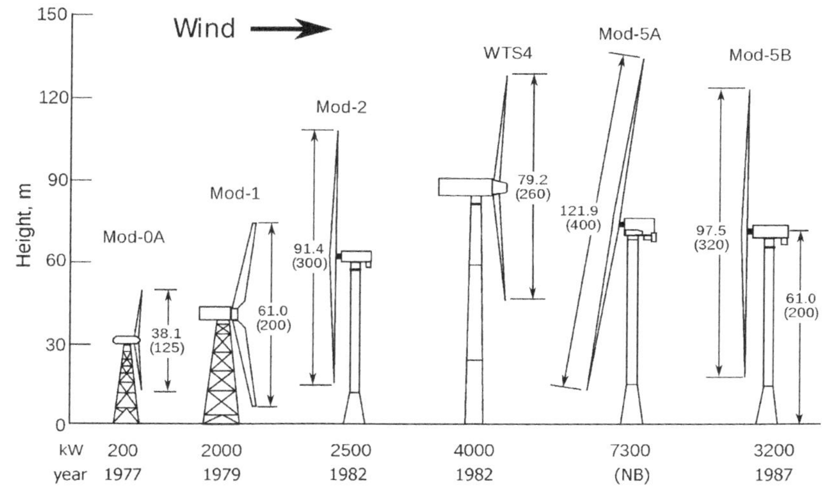

Figure 1.13 Schematic diagram to scale of NASA Mod-series wind turbines. Rotor diameters and hub heights given in meters (feet), rated capacity in kilowatts (kW), and year of first installation. NB = not built. NASA image obtained and adapted from Wikimedia Commons https://commons.wikimedia.org/

Figure 1.14 NASA Mod_0 wind turbine at Plum Brook Station, Sandusky, Ohio (1976). Downwind, two-bladed rotor was 38 m in diameter, hub height 30 m, and rated output 100 kW. Note the person at lower left for scale. NASA on The Commons/Wikimedia Commons/Public Domain.

Figure 1.15 U.S. Windpower turbines at Paterson Pass in the Altamont Pass wind-energy complex as seen in 2012. The turbines have three fiberglass blades with variable pitch mounted downwind from the tower. Rotor diameter is 17 m with nominal output of 100 kW (Musgrove 2010).

farm at Crotched Mountain, New Hampshire, quickly realized the potential tax and sales advantages in California. The company began leasing land at Altamont Pass and erected the first 100 wind turbines in 1981 (Fig 1.15). Thus began the California wind boom.

1.4.2 Developments in the 1980s

Many other companies quickly followed in California with leasing land, obtaining zoning approval, contracting with local utilities, raising financing, and erecting wind turbines by the hundreds and then thousands in the early 1980s. For a good return on investment, two factors were critical—reliable turbines and a windy location (Musgrove 2010). Three mountain passes well known for strong wind became the primary locales for California wind-farm development, namely Altamont Pass situated on ridges of the Diablo Range east of San Francisco Bay, Tehachapi Pass located between the southern end of the Sierra Nevada Mountains and the Tehachapi Mountains, and San Gorgonio Pass situated close to Palm Springs and next to the San Jacinto Mountains.

Within five years, more than 12,000 wind turbines had been installed in California representing more than 30 different models (Musgrove 2010). U.S. Windpower turbines proved generally reliable; however, most other turbines of American design and manufacture suffered widespread failures. Beginning in 1983, California wind-energy developers increasingly sought Danish turbines that already had established a reputation for robust, reliable service. Wind turbines rapidly covered hill sides and ridge tops, like mushrooms sprouting after a spring rain (Fig. 1.16).

Figure 1.16 Overview of the Patterson Pass wind farm which was commissioned in 1985 with several hundred Bonus and Nordtank turbines, as seen in 2012. Altamont Pass vicinity, east of Livermore, California.

Lattice steel towers were the norm at the beginning of the 1980s decade. Nordtank, a Danish company that made tank vehicles, introduced welded-steel tubular towers for wind turbines in 1980 (Madsen 2009). Within a few years, other turbine makers had adopted this type of tower because it was visually more acceptable to most people (Musgrove 2010). Peter Mørup, the head of Nordtank manufacturing, left the company in 1983 and co-founded Micon with his brother.

Danregn Vindkraft was founded in 1980 by Peter Stubkjær Sørensen and Egon Kristensen. The early Danregn turbine was inspired by the Smedemester design and had many similarities. The company name soon was changed to Bonus. By the mid-1980s Vestas, Bonus, Nordtank, and Micon had become the leading Danish manufacturers of wind turbines that had gained a reputation for dependable performance (Musgrove 2010). It should be noted these companies had their origins in agricultural and transportation manufacturing, wherein reliable equipment is paramount, and they drew on the shared, grass-roots efforts from the 1970s.

The Nordic Folkecenter for Renewable Energy was founded in 1983 by Maegaard near Hurup, Denmark. The center is a museum for early wind turbines and provides technology transfer and pilot projects in many countries around the world. By this time more than 20 Danish companies produced wind turbines, and government subsidies were in place for purchase of turbines. Thus, the grass-roots era of wind-turbine development had come effectively to its end by the early 1980s.

The United States, however, failed to renew tax incentives for wind energy in 1986, which led to financial collapse for the Danish turbine industry the following year. Most companies (except Bonus) went into bankruptcy. All were reorganized, downsized, and managed to continue. The Danish government encouraged consolidating many small suppliers into larger integrated companies that would compete on the international market. The convergence of innovation, manufacturing, and public policy that resulted in Danish wind-energy success during the 1970–1980s is an example of *bricolage*, a French term that refers to resourcefulness in the creation of a new technology from diverse sources and available resources (Nielsen 2009).

The environmental movement had gained considerable momentum and widespread acceptance during the 1970s. Many of the strongest supporters followed the *small is beautiful* philosophy (Graetz 2011). This movement was especially opposed to nuclear power and also to large hydropower projects. Wind power remained the most promising alternative means for generating electricity; however, environmentalists were not particularly happy with the prospects for expanded wind energy. According to Graetz (2011, p. 123), *wind turbines are neither small nor beautiful*.

Lucrative tax credits in the United States attracted all types of investors seeking financial shelters; wind farms were regarded by some as "tax farms" (Graetz 2011). Along with serious investors came fraudulent and unscrupulous scoundrels. The most notorious case involved International Dynergy, which equipped a wind farm in southern California with used helicopter blades attached to broken turbines in an effort to fool tax appraisers.

Under the Reagan administration, federal tax credits expired at the end of 1985, and California tax credits ended a year later. Meanwhile oil prices declined sharply in 1986, and concerns arose about bird mortality, particularly at Altamont Pass. Public desire for wind energy diminished, and permitting of new wind farms became more difficult (Musgrove 2010). These factors in combination busted the California wind boom. Still, California utilities had long-term contracts to purchase electricity from wind farms, which allowed the industry to continue, albeit at a much slower pace than prior to 1986. U.S. Windpower survived on electricity sales, and in 1988 it was reorganized as Kenetech that diversified and expanded into new wind-energy markets.

James Hansen, a NASA climatologist, is often considered the father of global warming. In 1987, he testified before the U.S. Congress concerning the role of atmospheric carbon dioxide and other greenhouse gases for global warming (Graetz 2011). He warned that climatic change could dramatically impact global environments and the world's human population. Although a scientific and political consensus was still in the future, the potential impact of burning fossil fuels was clear, and the need to develop renewable, non-polluting sources of energy became more urgent.

1.4.3 Developments in the 1990s

By 1990, the Danish concept had proven superior for routine, dependable generation of electricity from the wind under wide-ranging circumstances and conditions. Efficient, low-cost manufacturing techniques had been worked out, so that Denmark was well positioned to dominate the wind-energy market during the 1990s. One key to improved economics was increasingly large turbines, which had reached 25–27 m in rotor diameter and 150–250 kW output. By 1996, Vestas had achieved 20% of the world market with Bonus, Nordtank, and Micon each around 10% for a total of half the global market in wind turbines (Musgrove 2010). The latter two companies merged in 1997 to form NEG Micon.

Already the Danish government began to recognize limits for land-based wind farms. The world's first offshore wind farm was constructed at Vindeby in 1991 in shallow seas of relatively protected waters between islands of southern Denmark (Fig. 1.17). Offshore wind farms, although more expensive to construct, produce substantially more electricity than land-based wind farms because of stronger and more consistent wind. Most of the land area of Denmark, for example, has average wind speeds at 100 m height of 7–9 m/s; whereas, much of the surrounding sea is 9–10 m/s (GWA 2023). By the end of the decade, Denmark had formulated plans to build large offshore wind farms.

Figure 1.17 Vindeby turbine preserved and displayed in the Danish Energy Museum. Blades are 17 m long, the tower originally stood 33 m tall, and generating capacity was 450 kW. Note the brake flaps in the blade tips. From the world's first offshore wind farm erected in 1991 and taken offline in 2017.

Figure 1.18 Pair of Tacke turbines at Swarzewo, northern Poland, as seen in 1998. Tacke Windtechnik was the second leading German manufacturer of wind turbines in the 1990s.

Europe was poised for rapid expansion of wind energy during the 1990s. Germany and Spain enacted feed-in tariff laws for sale of renewable energy to large utilities, which led to exponential growth of their wind industries (Musgrove 2010). Gamesa and Acciona emerged as major turbine manufacturers in Spain; Gamesa entered the wind industry in 1993 in partnership with Vestas. In Germany, most turbines were installed near the North Sea and Baltic Sea coasts. As in Denmark, turbine dimensions grew steadily, reaching megawatt size by the end of the decade. The main German turbine makers were Enercon and Tacke (Fig. 1.18). Tacke, however, fell on economic difficulties and declared bankruptcy in 1997. It was bought by Enron Wind and subsequently acquired by General Electric in 2002.

The feed-in tariff laws in Denmark, Germany, and Spain had been highly successful for stimulating the wind industries in those countries. The United Kingdom, in contrast, took a different approach that proved quite disappointing (Musgrove 2010). The Non-Fossil Fuel Obligation (NFFO) was devised during 1989–1990 primarily to support the higher cost of producing nuclear energy by placing a levy on electricity generated with fossil fuels. Only a small portion of proposed British wind farms were approved and eventually constructed.

By the end of 2000, the total U.K. capacity was just 410 MW compared with 2,300 MW in Denmark, 2,400 MW in Spain, and 6,100 MW in Germany (Musgrove 2010). By the late 1990s, the lack of NFFO success was evident, and the government moved to replace it. The result was the Renewable Obligation (RO), a complex process that required utilities to sell electricity generated from renewable energy resources; it began early in the twenty-first century.

Both China and India have quite large potentials for wind energy. India began early development in the mid-1980s with financial support from Denmark. In the mid-1990s, both national and state governments provided generous financial incentives for wind energy, which exceeded a total capacity of 1,000 MW in 1999. The Indian market stimulated domestic production, and Suzlon emerged as a major builder of wind turbines. China likewise began limited wind-energy development in the late 1980s also with financial support from Denmark. However, China's pace of wind-energy growth remained slow during the 1990s. Goldwind produced its first Chinese turbine in 1998 with German technical help.

Following the collapse of the California wind boom, growth of wind energy stagnated in the United States. Between 1990 and 1997 installed capacity increased by a trivial 90 MW (Musgrove 2010). The federal Production Tax Credit (PTC) of 1992 was an attempt to increase use of renewable energy technologies; financial support was linked to a project's long-term energy output rather than its capital cost. The PTC revived the U.S. wind industry, which began a modest recovery during the decade. The most successful venture during this period was the wind farm developed on Buffalo Ridge in southwestern Minnesota.

Buffalo Ridge

Buffalo Ridge is the most conspicuous drainage divide in the north-central United States. The ridge is a glacial landform, known as the Bemis Moraine, that accumulated along the western edge of the Des Moines ice lobe about 14,000 years ago. The Bemis Moraine extends from eastern South Dakota, across southwestern Minnesota, and into northwestern Iowa. Buffalo Ridge marks the eastern margin of the Prairie Coteau, a glacial upland that drains into the Missouri River basin. Buffalo Ridge is the divide, east of which drainage heads into the Minnesota River and eventually Mississippi River basin. The ridge exceeds 600 m elevation toward its northern end at Summit, South Dakota, some 300 m above the Minnesota River valley to the east.

(*Continued*)

Buffalo Ridge (Continued)

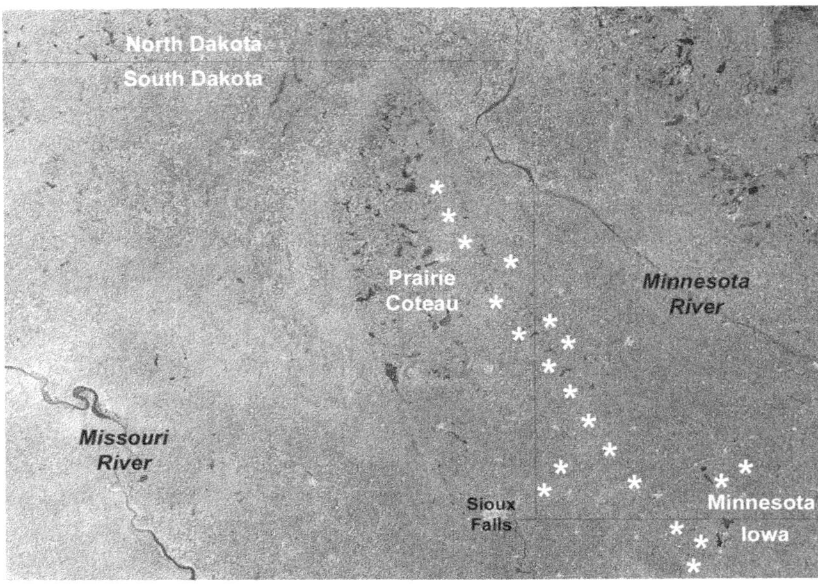

Satellite image of the Great Plains region in the north-central United States. Numerous wind farms (*) are positioned along Buffalo Ridge on the eastern margin of the Prairie Coteau upland in South Dakota, Minnesota, and Iowa (Hoen et al. 2018–2025). MODIS natural-color image, Terra satellite, September 3, 2002. NASA / Public Domain.

This prominent drainage divide was an early target for wind-energy development during the 1990s. When Northern States Power requested to expand its nuclear-waste storage, Minnesota required that it add 425 MW of wind power to its generating inventory. Kenetech turbines were used in the first phase of development in 1994. These were followed by Zond turbines in the second phase completed in 1998. Zond was bought by Enron and renamed Enron Wind. Meanwhile Kenetech, once the world's largest manufacturer of wind turbines, suffered continuing engineering problems and went bankrupt in 1996.

Wind farms have proliferated along Buffalo Ridge and the Prairie Coteau since then from near Summit, South Dakota, to Worthington, Minnesota, a distance of more than 200 km, and continuing southward into Iowa. More than 2,700 turbines are arrayed in these wind farms in South Dakota and Minnesota (Hoen et al. 2018–2025). Multiple generations of operational wind farms and turbines range in age from the late 1990s to early 2020s, including the original Zond/Enron turbines from 1998. These turbines demonstrate the trend through time for larger sizes and capacities. Meanwhile, some of the oldest and smallest turbines have been removed.

Buffalo Ridge (Continued)

Kenetech KVS-33 wind turbines on Buffalo Ridge near Lake Benton, southwestern Minnesota, as seen in 1996. Rotor diameter is 33 m, and output is 300–360 kW. Between 1992 and 1996, about 700 of these turbines were installed at sites across the United States and in Costa Rica (Carlin et al. 2001).

Older and newer turbines on Buffalo Ridge near Lake Benton. (Left) Vestas 0.66 MW turbines in the Ruthton Wind Farm date from 2001. Rotor diameter is 47 m, and total height is 88.4 m (290 feet). (Right) GE Wind 2.82 MW turbines of the Buffalo Ridge wind farm were installed in 2022 surrounding the Ruthton vicinity. Rotor diameter is 127 m, and total height is 152.4 m (500 feet).

Besides Minnesota, several other states began to require utilities to provide electricity generated with renewable energy through some form of Renewables Portfolio Standard (RPS). This came about mainly because of growing concerns about greenhouse gases and climate change, and development of new wind-energy facilities began to pick up in the late 1990s. But the PTC was allowed to lapse in 1999, which once again stalled wind-power developments. The PTC was soon renewed, but did not have any practical effect until 2001. This stop-and-go cycle of financial incentives for wind energy was symptomatic and highlights the lack of a coherent, long-term energy policy at the federal level, a problem that still continues.

Part of the problem was strong opposition to wind energy by those with vested interests in fossil fuels as energy sources. For example, in 1999 the American Association of Petroleum Geologists (AAPG) rejected the likelihood that human activity, namely burning fossil fuels, has any influence on global climate change (Bean 2012). AAPG has since softened its position on this subject, but opposition to development of wind energy remains strong even today among many in the fossil-fuel industries.

1.5 Wind Energy—Early Twenty-First Century

In 1990, the global wind-power capacity was just 2,000 MW. Ten years later, total capacity had reached 17,700 MW (Musgrove 2010). Still, wind energy was a niche enterprise at the beginning of the twenty-first century. Since then, wind energy has expanded dramatically around the world for varied reasons. Many countries provided financial incentives to increase the domestic production of renewable and non-polluting energy in order to lessen their use of fossil fuels and nuclear energy as well as to decrease their dependence on unreliable foreign energy sources.

The use of wind power for generating electrical energy has grown phenomenally during the past quarter century. Total installed wind-power capacity worldwide by the end of 2023 had exceeded 1,000 gigawatts (GW) led by China (441.9 GW) and the United States (148.0 GW), which together accounted for more than half of global wind-generating capacity. Germany and India were the next leading countries followed by Spain and the United Kingdom (Table 1.3). Maegaard et al. (2013) and Aber et al. (2015) presented summaries for history and status of wind energy in many countries around the world.

In the United States, renewal of the PTC combined with RPS mandates in more than 25 states led to rapid increases in wind energy starting in 2001 (Fig. 1.19). The goal to control greenhouse gases and climate change came to the forefront of energy and policy debates, particularly under the administrations of President Obama and President Biden. In particular, the Inflation Reduction Act of 2022 addressed energy security and climate change, although the issue remains a low priority for many citizens.

General Electric (GE) entered the wind industry with purchase of Enron Wind's manufacturing assets in 2002 to form GE Wind Energy (Musgrove 2010). It developed

Table 1.3 Top 12 countries for wind-energy generating capacity including both onshore and offshore installations.

Country	Installed (GW)	Country	Installed (GW)
China	441.9	Brazil	29.1
United States	148.0	France	20.8
Germany	69.5	Canada	17.0
India	44.7	Sweden	16.3
Spain	31.0	Turkey	11.7
United Kingdom	30.1	Australia	12.9

Rounded values in gigawatts (GW) as of end 2023. Adapted from IRENA (2024).

Figure 1.19 Gray County Wind Farm, built in 2001, was the first large array of turbines in Kansas. It has 170 Vestas V47 turbines with a combined capacity of 112 MW. These turbines are relatively small and closely spaced compared with newer models. Kite aerial photograph (2006).

an earlier Tacke design for a 1.5-MW turbine that was installed at many wind farms across the country (Fig. 1.20). GE became the dominant maker of wind turbines in the United States. In 2024, GE Vernova was spun off from its parent, and the new company deals primarily with renewable energy worldwide. Its headquarters is located in Massachusetts (GE Vernova 2024).

Vestas and Siemens also continued to supply the U.S. market as well as Gamesa, Mitsubishi, and Suzlon. The sight of wind-turbine components in transit on special truck trailers and railway carriages became commonplace, especially in the Great Plains region (Fig. 1.21). New wind farms proliferated, and existing wind farms were expanded. International companies built manufacturing plants in the United States to supply the rapidly growing market—Vestas in Colorado, and Siemens in Kansas and Iowa.

Figure 1.20 Spearville Project, built in 2006, is equipped with GE Wind 1.5 MW turbines amid agricultural fields. These GE Wind turbines were made in the United States; they are descendants of the Tacke turbine made in Germany during the 1990s. Kite aerial photograph with D. Leiker and C. Unruh (2007).

Wind-energy capacity grew quickly in both China and India, and China surpassed the United States to become the world's leading producer. China has expanded aggressively for offshore wind farms, which now account for more than 8% of total Chinese wind-energy capacity, but India has yet to build any offshore

(A)

Figure 1.21 Turbine-blade logistics in Kansas. (A) Highway transport by oversized tractor–trailer truck; blades are typically about 50–60 m long, (B) temporary storage in the BNSF Railway depot at Emporia. Vertical view of Gamesa blades in tight-packing arrangement. Service truck shows size of blades. Kite aerial photograph with D. Leiker.

Figure 1.21 (Continued)

(B)

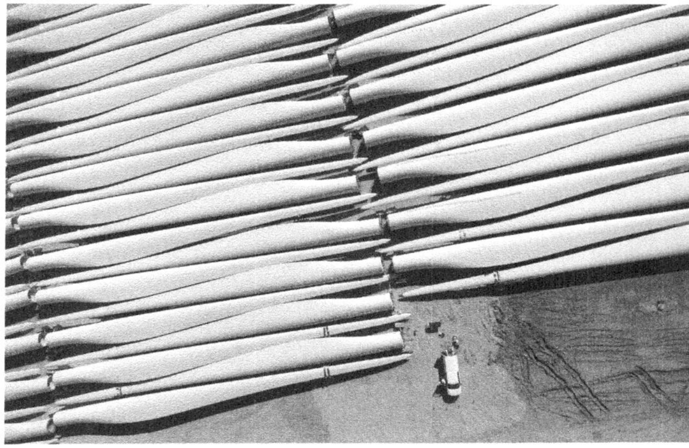

wind farms (GWEC 2024a). Six of the ten largest wind-turbine manufacturing companies are located in China (Table 1.4).

Germany maintained its dominance of European wind energy, and the United Kingdom expanded considerably with offshore wind farms, which now make up nearly half of its total capacity (GWEC 2022). Siemens (a German company) acquired Bonus Energy in 2004, which was renamed as Siemens Wind Power. Gamesa and

Table 1.4 Top 10 wind-turbine companies generating capacity supplied for 2023.

Company	Country	Capacity
Goldwind	China	16.7
Envision	China	15.9
Vestas	Denmark	12.7
Windley	China	10.5
Mingyang	China	10.2
Siemens Gamesa	Spain	7.7
GE Vernova	United States	7.6
SANY	China	7.5
Nordex group	Germany	6.9
Dongfang	China	5.6

Six of the ten companies come from China, which is also the largest consumer for wind turbines. Values given in gigawatts (GW) rounded.
Adapted from GWEC (2024b).

Figure 1.22 EHN turbines harvest wind energy on a prominent ridge near Pamplona, northern Spain. Photo courtesy of J.T. Aber (2012).

Figure 1.23 Middelgrunden wind farm in the Øresund strait between Denmark and Sweden. A single, curved alignment; commissioned in 2001 with 20 Bonus 2.0 MW turbines. This wind farm is highly visible to the public given its location near Copenhagen, Denmark. Arnoldius/Wikimedia Commons/CC BY-SA 3.0.

Siemens merged in 2017 to form Siemens Gamesa Renewable Energy (SGRE) with headquarters in Spain.

Acciona Windpower built its first wind farm in Spain in 1994 and since expanded rapidly. In 2005, Acciona bought full ownership of EHN, a leading Spanish developer of wind energy (Fig. 1.22). Nordex was founded in Denmark in 1985, part of the Danish revolution in wind energy (Nielsen 2009), but then Nordex moved to Germany in 1992. Nordex and Acciona merged in 2016 with headquarters in Germany.

During the first decade of this century, Danish development moved increasingly toward large wind farms, both on land and offshore into the surrounding seas (Fig. 1.23). Offshore wind turbines typically produce about 50% more electricity than comparable onshore turbines, which justifies the higher cost of building offshore wind farms (Musgrove 2010). As of 2022, offshore wind turbines provided 2.3 GW of capacity, approximately one-third of Denmark's total wind power.

Onshore, meanwhile, the government offered incentives to replace many smaller, older turbines with fewer, larger turbines. Denmark and its neighbor Sweden continue to lead the world in per-capita wind-generated electricity with nearly 3.3 kWh per person (Ember 2024), and Vestas remains among the world's largest wind-energy companies (Table 1.4). As the foregoing discussion illustrates, the early twenty-first century was a period of tremendous expansion of wind energy worldwide as well as an interval of mergers and consolidation within the wind industry.

1.6 Summary

People have utilized the wind to power various types of devices and vessels since the dawn of civilization. The classic European windmill first appeared in the twelfth century in France and England and spread rapidly thereafter. The era of such windmills lasted until the early twentieth century, some eight centuries. In the middle nineteenth century, windmill innovation took a completely different direction in the United States. The continuing success of the American windmill is due to several factors—ease of manufacturing, shipping, assembly, routine maintenance, and autonomous operation.

Generating electricity with windmills began in the late nineteenth century in several locations. The most important developments were achieved in Denmark by Poul la Cour in the 1890s and beginning of the twentieth century. His innovations may be traced forward into the modern wind industry. Johannes Juul's wind turbine at Gedser, Denmark, and Ulrich Hütter's fiberglass blades are particularly noteworthy for demonstrating many technical aspects that would become industry standards later in the twentieth century.

As cheap oil and nuclear energy became widely available in the 1960s, the perceived need for wind energy reached a low ebb, and most wind-energy projects were shut down. At this time, the public and policy makers believed that cheap energy was unlimited and without environmental consequences, but these assumptions changed quickly with a series of oil crises and nuclear disasters beginning in the 1970s.

Among the most important wind-energy developments during the 1970s and 1980s were those in Denmark and the United States. By the end of the 1970s, Denmark had moved through an extremely rapid and innovative phase of wind-energy development, which led to the so-called Danish concept for modern wind turbines. A modest wind-energy program in the United States was initiated in the 1970s with a series of increasingly large Mod turbines, and a combination of federal and state tax incentives set off the California wind boom in the early 1980s.

The California wind boom crashed when tax credits expired in 1985 (federal) and 1986 (state). Concerns about greenhouse gas emissions from burning fossil fuels began to emerge as part of the debate about energy resources. During the 1990s, one key to improved economics was increasingly large turbines. Danish companies had a combined share of roughly half of the world market for wind turbines, and Denmark began to experiment with offshore development of wind farms.

The early twenty-first century was a period of tremendous expansion of wind energy worldwide as well as a time of mergers and consolidation within the wind industry. Many countries provided financial incentives, and by 2023 global wind-energy capacity exceeded 1,000 GW. China, United States, Germany, India, Spain, Brazil, and the United Kingdom emerged as the largest producers.

Offshore wind farms were pioneered by Denmark and have expanded considerably particularly for China and the United Kingdom. Denmark and Sweden continue to lead the world in per-capita wind-energy production, and Vestas remains among the largest company for manufacturing wind turbines. Thus, wind power has emerged from unlikely prospects based on the Danish concept to become a major source for renewable, non-polluting energy worldwide in the twenty-first century.

References

Aber, J.S., Aber, S.W. and Pavri, F. 2015. Windscapes: A global perspective on wind power. Multi-Science Publ., UK, 245 p.

Autocar 1961. Correspondence (photograph). The Autocar, May 12, 1961, p. 758.

Baker, T.L. 1985. A field guide to American windmills. University of Oklahoma Press, Norman, Oklahoma, 516 p.

Bean, D. 2012. Drilling into data: Oilman says the facts are clear that CO_2 is a threat. *Binghamton University Magazine* 8(3), p. 16–21.

Beedell, S. 1975. Windmills. C. Scribner's Sons, New York, 143 p.

Carlin, P.W., Laxson, A.S. and Muljadi, E.B. 2001. The history and state of the art of variable-speed wind turbine technology. National Renewable Energy Laboratory, Technical Report 500-28607.

Cavey, J.-L. 2024. The Nogent Le Roi wind turbine (France) 1955-1966 EDF – BEST – Romani. http://eolienne.cavey.org/en/index_en.php

Christensen, B. 2008. Danske og underlandske forløbere for FLS-"Aeromotorerne". Kapitler af Vindkraftens Historie i Danmark, 4. årgang, p. 10.

Christensen, B. 2009. Danish windmill production – Before and after Poul la Cour – And the first "Golden Age" of wind power 1900-1920. In: B. Christensen (ed.), Wind power the Danish way: From Poul la Cour to modern wind turbines, p. 24–31. The Poul la Cour Foundation, Askov, Denmark, 88 p.

Christensen, B. 2016. Den rigtige mølle – men på det forkerte tidspunkt. Kapitler af vindkraftens historie i Danmark. 11. årgang, p. 22–23. Energimuseet – Nordiske Folkecenter for Vedvarende Energi – Poul la Cour Museet – Danmarks Vindkrafthistoriske Samling.

Christensen, B. 2017. Gårdmøllerne – Den første Danske Vindmølle-Eventyr. Energimuseet – Nordiske Folkecenter for Vedvarende Energi – Poul la Cour Museet – Danmarks Vindkrafthistoriske Samling. Øko Tryk, Skjern, Denmark, 72 p.

Christensen, B. and Thorndahl, J. 2012. Fra husmøller til havmøller: Vindkraft i Danmark i 150 år. Energimuseet, Nordisk Folkecenter for Vedvarende Energi, Poul la Cour Foundation, Danmarks Vindkrafthistoriske Samling, 56 p.

Ember 2024. Electricity generation from wind power per person. Ember and Energy Institute – Statistical Review of World Energy (2024). Population based on various sources (2023) – With major processing by Our World in Data. https://ourworldindata.org/grapher/wind-electricity-per-capita

GE Vernova 2024. A unified and powerful force. GE Vernova. https://www.gevernova.com/about

Graetz, M.J. 2011. The end of energy: The unmaking of America's environment, security, and independence. MIT Press, Cambridge, Massachusetts, 369 p.

Grove-Nielsen, E. 2023. Winds of change: Stories: Winds of a dawning wind power industry. Norkisk AeroForm ApS. https://windsofchange.dk/

GWA 2023. Global wind atlas 3.3. https://globalwindatlas.info/en

GWEC 2024a. GWEC global wind report 2024. Global Wind Energy Council. https://img.saurenergy.com/2024/05/gwr-2024_digital-version_final-1-compressed.pdf

GWEC 2024b. Wind turbine manufacturers see record year driven by growth in home markets. Global Wind Energy Council. https://www.evwind.es/2024/05/09/wind-turbine-manufacturers-see-record-year-driven-by-growth-in-home-markets/98394

Hoen, B.D., Diffendorfer, J.E., Rand, J.T., Kramer, L.A., Garrity, C.P. and Hunt, H.E. 2018. United States Wind Turbine Database V8.0 (February 25, 2025). U.S. Geological Survey, American Clean Power Association, and Lawrence Berkeley National Laboratory data release. https://energy.usgs.gov/uswtdb/ or https://doi.org/10.5066/F7TX3DN0

IRENA 2024. Renewable energy statistics 2024. International Renewable Energy Agency. https://www.irena.org/-/media/Files/IRENA/Agency/Publication/2024/Jul/IRENA_Renewable_Energy_Statistics_2024.pdf

Jacobs Wind Electric 2019. Jacobs wind electric: About us. http://www.jacobswind.net/

KHS 2014. Currie windmill. Kansapedia, Kansas Historical Society. https://www.kshs.org/kansapedia/currie-windmill/10150

Madsen, B.T. 2009. Public initiatives and industrial development after 1979. In: B. Christensen (ed.), Wind power the Danish way: From Poul la Cour to modern wind turbines, p. 52–59. The Poul la Cour Foundation, Askov, Denmark, 88 p.

Maegaard, P. 2009. The new wind power pioneers and the emergence of the modern wind industry. In: B. Christensen (ed.), Wind power the Danish way: From Poul la Cour to modern wind turbines, p. 46–51. The Poul la Cour Foundation, Askov, Denmark, 88 p.

Maegaard, P., Krenz, A. and Palz, W. (eds.). 2013. Wind power for the world: International reviews and developments. Jenny Stanford Publishing, 730 p.

Musgrove, P. 2010. Wind power. Cambridge University Press, 323 p.
Nielsen, K.H. 2009. International perspectives on the history of Danish wind power. In: B. Christensen (ed.), Wind power the Danish way: From Poul la Cour to modern wind turbines, p. 60–65. The Poul la Cour Foundation, Askov, Denmark, 88 p.
Nissen, P.-O. 2007. Det teknologiske og oplysende kraftcentrum i Askov – fra knaldluft over aerodynamik til udbredelse a kendskabet til elektricitet, bygning at vindkraftværker samt uddanelse af tilhørende personale. Kapitler af Vindkraftens Historie i Danmark, 3. årgang, p. 9–15. Energimuseet – Nordiske Folkecenter for Vedvarende Energi – Poul la Cour Museet – Danmarks Vindkrafthistoriske Samling. https://povlonis.dk/wp-content/uploads/2015/05/askov-artikel8fd3.pdf
Nissen, P.-O. 2009. The scientist, inventor and teacher Poul la Cour. In: B. Christensen (ed.), Wind power the Danish way: From Poul la Cour to modern wind turbines, p. 6–11. The Poul la Cour Foundation, Askov, Denmark, 88 p.
Price, T.J. 2005. James Blyth: Britain's first modern wind power pioneer. *Wind Engineering* 29(3), p. 191–200. https://doi.org/10.1260/030952405774354921
Quistgaard, T. 2009. The experimental windmills at Askov 1891-1903. In: B. Christensen (ed.), Wind power the Danish way: From Poul la Cour to modern wind turbines, p. 12–17. The Poul la Cour Foundation, Askov, Denmark, 88 p.
Ragheb, M. 2023. Historical wind generators machines. Wind Power Systems. http://mragheb.com/NPRE 475 Wind Power Systems/Historical Wind Generators Machines.pdf
Thorndahl, J. 2005. Gedsermøllen – den første moderne vindmølle. Skrifter fra Elmuseet nr. 15, Tange, Denmark.
Thorndahl, J. 2009. Johannes Juul and the birth of modern wind turbines. In: B. Christensen (ed.), Wind power the Danish way: From Poul la Cour to modern wind turbines, p. 40–45. The Poul la Cour Foundation, Askov, Denmark, 88 p.
Thorndahl, J. 2011. En vindmølle i baghaven: Om vindkraft pioneren Christian Riisager. Kapitler af Vindkraftens Historie i Danmark, 7. årgang, p. 2–9. Energimuseet – Nordiske Folkecenter for Vedvarende Energi – Poul la Cour Museet – Danmarks Vindkrafthistoriske Samling.
Thorndahl, J. and Christensen, B. 2009. Time for survival and development 1920-1945. In: B. Christensen (ed.), Wind power the Danish way: From Poul la Cour to modern wind turbines, p. 32–39. The Poul la Cour Foundation, Askov, Denmark, 88 p.

2

Windscape Methodology

A map is a graphic representation of spatial concepts; it is but another method of communication. That method is universal—it can be understood and appreciated by all (Merriam 1996, p. 95).

2.1 What Is the Windscape?

How can we understand the power in the wind in relation to the greater environment and human perspective? Civil engineers focus on the bedrock, soils, drainage, and other conditions that influence construction of wind turbines and related wind-farm infrastructure. A meteorologist considers micro-climate and long-term wind conditions paramount factors. Biologists look at the habitat and wildlife with an emphasis on any rare or endangered species living in or migrating across the site.

Policy makers and public officials consider the right mix of incentives, financial or otherwise, in spurring on investment in wind-energy production. A local landowner might consider the economic consequences—both financial incentives as well as impacts on existing land use and value. Visibility, noise, and health issues could be most important to casual visitors as well as nearby permanent residents. Local land-use planners and town councils may enact ordinances and zoning requirements to minimize the human and environmentally related impacts of large wind-energy infrastructure facilities.

The term windscape is a common word used in a wide variety of contexts. Here we apply the term to the combination of natural conditions and human exploitation of a site for wind energy (Aber et al. 2015). This is similar to Brittan's (2002) *wind energy landscape,* which includes wind-power technology as well as siting issues, ownership, operation, and human perceptions of wind farms.

Wind Energy Landscape: Principles, Techniques, History, First Edition. James S. Aber, Susan E. W. Aber, and Firooza Pavri.

Published 2026 by John Wiley & Sons, Inc.

The windscape begins with the land or sea floor on which wind turbines are erected. Wind, of course, is the next consideration in terms of both regional wind patterns as well as local conditions that influence wind speed and direction. Vegetation cover and human land use are further factors that interact with near-surface wind. The existing grid system and need for electricity likewise determine the locations of potential wind farms. Environmental factors, history, and cultural expectations might dictate what types of land use could or may not be acceptable. All of these issues are relevant to understanding the windscape.

2.2 Geospatial Analysis

Means of investigating and analyzing the windscape vary widely for particular disciplines and purposes. In this book, we emphasize maps and geospatial analysis, also known as geographic information systems or science (GIS). Geospatial analysis emerged in the late twentieth century as a means to integrate physical, biological, and cultural aspects for all manner of geographic investigations and now has become a routine method for analyzing diverse environmental and human applications (e.g. Burrough 1986; Aber and Aber 2017). GIS combines advanced computer mapping, remote sensing, statistical analysis, and visualization techniques to portray relationships between what, where, and when events and conditions happen. In other words, GIS helps us to understand the connections between the natural world and impacts of human activities.

GIS combines datasets collected on the ground by direct measurement and surveying with aerial observations, satellite imagery, and other forms of remote sensing. Scale and resolution are key aspects for geographic information; they determine the smallest objects or features that may be positively identified or mapped from available data. In general landscape and ecological studies, three primary levels or scales are commonly recognized (Aber et al. 2012).

- Micro-landscape—Individual features—hill, hummock, swale, crevasse, boulder, tree, animal, etc. Features of the size are easily seen and measured directly in the field by a person on the ground. Spatial resolution <1 m; map scale greater than 1:1,000.
- Meso-landscape—Assemblages of related or closely connected features—coastal, bog or delta complex, forest tract, alpine glacier, volcano, coral reef, large animal herd, etc. Areas too large to see completely from any one location on the ground. Spatial resolution 1–100 m; map scale 1:1,000 to 1:100,000.
- Macro-landscape—Collections of several environmental types spanning a large geographic area or range in altitude—continental shelf, tropical rainforest, mountain range, ice sheet, etc. Areas that require regional mapping to portray. Spatial resolution >100 m; map scale less than 1:100,000.

We herein utilize those methods that reveal the physical, biological, and cultural aspects of windscapes, namely GIS and remote sensing. These techniques must be combined with conventional field observations in order to provide ground truth for validating, analyzing, and interpreting remotely sensed information. From the frog's-eye view on the ground to the vantage from outer space (Fig. 2.1), we now have the ability to observe the windscape from many heights and resolutions. It is this multi-level approach that yields comprehensive and convincing results.

(A)

(B)

Figure 2.1 Wind farms east of San Francisco Bay, California. (A) Patterson Pass wind farm with Bonus 1980s-vintage Danish turbines as seen in 2012. (B) Landsat 9 natural-color satellite image acquired December 10, 2021. Locations of wind farms (*) at Altamont Pass and Patterson Pass. USGS Landsat (2022)/USGS/Public Domain.

2.3 Remote Sensing

We may observe the windscape from above using aerial photography, satellite imagery, and other means of remote sensing. Information is conveyed by electromagnetic radiation that is reflected or emitted from objects on the ground or in the air, transmitted through the atmosphere, and collected by various types of cameras or detectors. In most cases, sunlight illuminates the scene and provides reflected visible and near-infrared wavelengths to generate images. For a full discussion of remote-sensing methods, the reader is advised to consult recent textbooks on the subject (e.g. Campbell et al. 2022).

The term resolution is used in widely different senses in various disciplines. In the general context of remote sensing, it refers to the smallest object or feature that may be identified by an experienced image interpreter. For digital imagery, the smallest object is a single cell or picture element (pixel) in the scene. Each pixel corresponds to an actual spot on the ground, and the linear size of this spot is the image spatial resolution. However, the identity of a single pixel rarely may be determined by itself. For positive identification, usually a group of four to nine pixels is necessary (Hall 1997).

The spatial resolution of an aerial photograph or satellite image is only one factor that comes into play for image interpretation. Other basic aspects of resolution are temporal and spectral characteristics. Additional factors include gray tone or color, shape, size and height, shadows, pattern and texture, and site context (based on Avery and Berlin 1992; Jensen 2007).

Each of these basic features is seldom recognized on its own. Rather it is the combination of visual clues that allows interpretation of objects depicted in aerial or space imagery. With these factors in mind, a perceptual measure of image quality or interpretability has been developed (Table 2.1). An image rating depends on the most difficult interpretation task that can be accomplished. Levels 1–3 represent the largest and most generalized features, which correspond to regional and macro-scales. The middle three interpretability classes (4–6) are at the meso/micro-levels, and the highest interpretability levels (7–9) represent the micro- and sub-micro-classes.

2.3.1 Macro-level Systems

Several satellite systems provide low-spatial resolution imagery in visible and infrared portions of the spectrum. These include the Advanced Very High Resolution Radiometer (AVHRR) carried onboard several NOAA weather satellites as well as the Moderate Resolution Imaging Spectroradiometer (MODIS) instrument on NASA's Terra and Aqua satellites. They have multiple functions for display and analysis of atmospheric conditions and land cover over broad areas (Fig. 2.2).

At the moderate-resolution level, Landsat is the longest running of all Earth-resources satellite systems beginning in 1972. The success of Landsat spawned

Table 2.1 The civilian U.S. National Imagery Interpretability Rating Scale (NIIRS) levels (0–9) and examples of interpretation tasks adapted for wind energy.

Rating level	Scale/ resolution	Interpretation tasks
0	—	Interpretability precluded by obscuration, degradation, or poor resolution
1	MACRO	Distinguish between major landuse classes—urban, agricultural, forest, etc. Detect medium-sized port facility Distinguish between runways and taxiways at a large airport Identify regional drainage patterns—dendritic, trellis, etc.
2	MACRO	Identify large (800 m diameter) center-pivot irrigated fields Detect large buildings—hospital, factory, etc. Identify road pattern—highway interchange, road network, etc. Detect ice-breaker tracks in sea ice; detect wake of large (100 m) ship
3	MACRO	Detect individual houses in residential neighborhoods Detect trains or strings of railroad cars on tracks Distinguish large wind farm, electrical substation Distinguish between natural forest and orchards
4	MESO	Identify farm buildings as bams, silos, or residences Count tracks along railroad right-of-way or in train yards Detect basketball court, tennis court, volleyball court in urban area Detect jeep trails through grassland
5	MESO	Recognize individual train cars by type—box, flat, tank, etc. Identify individual large wind turbine, power line structure, etc. Distinguish between deciduous and coniferous forest during leaf-off season Detect large animals (elephant, rhinoceros, bison) in grassland
6	MESO	Distinguish between row crops and small grains Identity passenger vehicles as sedans, station wagons, etc. Identity individual utility poles in residential neighborhoods Recognize main turbine components—nacelle, blade, etc.
7	MICRO	Identify individual mature plants in a field of known row crops Recognize individual railroad ties Detect individual steps on stairways Detect stumps and rocks in forest clearings or wet meadows
8	MICRO	Count individual baby farm animals—pigs, sheep, etc. Identify a survey benchmark set in a paved surface Recognize individual pine seedlings or individual water lilies on a pond Detect individual bolts of turbine tower base
9	MICRO	Identify individual seed heads on grain crop—wheat, oats, barley, etc. Recognize individual barbs on a barbed-wire fence Detect individual pebbles in turbine pad or access road Identify individual leaves on a tree

Adapted from Leachtenauer et al. (1997).

Figure 2.2 Fog blankets part of the San Joaquin Valley of central California and snow caps the Sierra Nevada Mountains in this MODIS natural-color image from November 2002. The locations of Altamont Pass (AP) and Tehachapi Pass (TP) wind-energy complexes are noted. SF—San Francisco, LA—Los Angeles. NASA/ Public Domain.

many similar Earth-resources satellites in the late twentieth century by the United States and several other countries (Lauer et al. 1997; Tatem et al. 2008). In 2008, the U.S. Geological Survey made archive Landsat datasets freely available to the public, which opened tremendous potential for long-term analysis of land-cover conditions.

Landsat imagery has proven of great value for all manner of environmental and human applications. We employ Landsat imagery to illustrate spatial and temporal characteristics of selected windscapes (see Fig. 2.1). Many band combinations are possible as well as multi-temporal composites (Pavri and Aber 2004). Macro-level systems are excellent for portraying regional topography, vegetation cover, climatic conditions, and human land use, but they are limited for showing details of wind farms or individual turbines.

2.3.2 Meso-level Systems

Conventional aerial photographs have been primary sources for mapping, surveying, and investigating all sorts of phenomena (e.g. Philipson 1997). These are most typically panchromatic (visible, gray tones), color, or color-infrared vertical pictures taken from several 1000 m above the surface at original (film) scales of 1:12,000 to 1:80,000 and spatial resolutions of 1–4 m. In many countries, useful aerial photographs date back 75 years or more and represent important historical images for documenting land cover and use. In some countries, however, aerial photography is classified for security reasons and is not available to the public or scientific researchers.

A common approach now is conversion of conventional aerial photographs into digital orthophoto quadrangles (DOQs). These images have been resampled to remove geometric distortions and to fit accurately onto a ground coordinate grid with high pixel resolution—usually 1 or 2 m. DOQs provide an excellent means to display wind farms and their surroundings as well as details for individual turbines and their arrangement (Fig. 2.3). A parallel trend is replacement of conventional aerial cameras employing large-format film by digital aerial cameras with similar spatial and spectral resolution.

Beginning in 1999 with Ikonos, commercial satellites have provided imagery with spatial resolution of 1–4 m and recently as small 30 cm (one foot). This spatial resolution rivals conventional airphotos and DOQs. These satellites also provide multispectral capability, typically in blue, green, red, and near-infrared bands, which is highly desirable for landscape imagery. The main drawback for this type of data is high commercial cost. *Google Earth* provides high-resolution imagery for many parts of the world, which is quite valuable for general geography; however, metadata are not given, which limits the usefulness of such images for detailed analysis purposes.

Figure 2.3 Portion of a digital orthophoto quadrangle (DOQ) derived from a conventional panchromatic aerial photograph showing part of the Gray County Wind Farm, Kansas (see Fig. 1.19). Individual wind turbines are clearly depicted by their distinctive shadows in this early spring view. Turbines are spaced approximately 160 m apart along field boundaries. Irrigation circles at upper right are about 0.8 km in diameter. Asterisk marks a small touristic kiosk next to U.S. highway 56. Image date March 20, 2002; USGS/Public Domain.

2.3.3 Micro-level Systems

To reach the highest levels of image interpretability (7–9), a spatial resolution of less than 30 cm is necessary. This may be achieved through small-format aerial photography (SFAP), which has been applied in recent years for documenting all types of natural and human resources (Aber et al. 2019). Manned or unmanned platforms include small aircraft, ultralights, balloons and blimps, kites, and unmanned aerial systems (UAS), commonly known as drones. These aircraft typically fly at heights of just a few 100 m. Many types of cameras may be utilized to acquire photographs in visible and near-infrared portions of the spectrum.

We fly tethered kites to collect aerial photographs from 50 to 150 m above the ground using light-weight radio-controlled digital camera rigs. Our method allows vertical as well as oblique imagery in all directions relative to the ground target, sun position, and shadows. This unique vantage, in proximity to large turbines, is especially suitable for oblique views of wind farms and their surroundings. We are able to acquire broad panoramas of the windscape as well as closer shots to illustrate prominent landforms and scenes that portray wind farms in their human context. A further benefit of kite aerial photography is a sensitive appreciation for near-surface wind behavior, particularly at the typical height that large turbines operate.

Kite Aerial Photography

Kite aerial photography (KAP) is a method to acquire low-height, high-resolution imagery including close-up views of wind turbines and overviews of wind farms. KAP is low in cost, highly mobile, quick to set up in the field, and relatively easy to operate. Legal restrictions for kite flying are few compared with UAS. In the United States, the maximum height for flying a tethered kite is normally 500 feet (152 m), which puts the camera at or above turbine hub height, while the kite may fly up to the same total height as many wind turbines.

Many other countries have similar height restrictions for tethered kites, for example 100 m in the Netherlands, but in the United Kingdom the normal height limit is only 60 m. Japan, on the other hand, has no height restriction for flying kites. At the opposite extreme, India is a country well known for strict security; in fact, taking any kind of aerial photograph is prohibited under nearly all circumstances. Any type of aerial photography, thus, needs to take into account local rules and national aviation laws (Aber et al. 2019).

KAP is a method we have used for depicting turbines and surrounding windscapes at a scale and vantage that most people are able to visualize and relate to ground features. As tethered aerial platforms, kites offer better control and safety compared with other methods, such as small UAS or manned ultralight aircraft, to cope with often turbulent conditions within or next to wind

Kite Aerial Photography (Continued)

farms. We have utilized kite aerial photography to document turbines and windscapes in the Great Plains of North America as well as in Denmark and Poland.

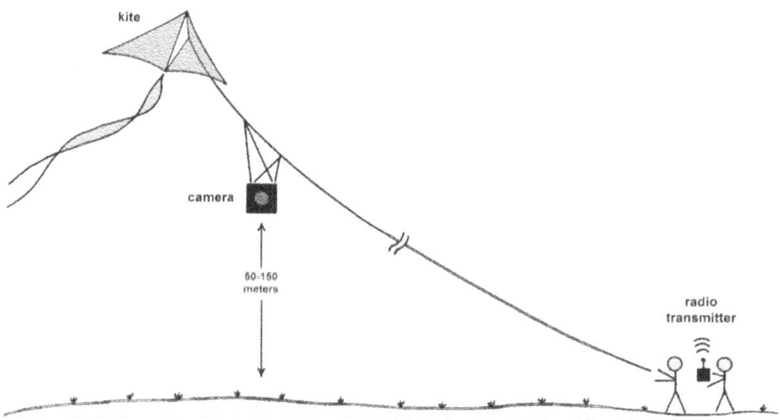

Cartoon showing the typical arrangement for kite aerial photography. The camera rig is attached to the kite line, and a radio transmitter on the ground controls operation of the camera rig. The kite line normally is anchored to a secure point on the ground. Aber et al. (2003).

Large rokkaku kite used most often by the authors for aerial photography. The kite measures 2.3 m tall and is flying here with a 6-m-long tube tail. Island of Bornholm, Denmark.

(Continued)

Kite Aerial Photography (Continued)

Bird's-eye view of Gamesa G90/2000 turbine; hub height is 80 m. Note the open aspect of the windscape west of Radwanki, Poland, with the forest margin in the distant background. Kite and camera flew in the downwind turbulence zone.

2.4 Maps and GIS

A map is a visual representation or scale model of spatial concepts. It is a means for conveying geographic information, namely those physical, biological, and cultural phenomena related to windscapes. Incorporated in a map is the understanding that it is a snapshot of an idea, a single picture, a selection of concepts from a constantly changing database of geographic information (Merriam 1996). GIS represents a major shift in the cartographic paradigm. In traditional (paper) cartography, the map was both the database and the display of geographic information. For GIS, however, data storage, analysis, and display are physically and conceptually separate aspects of handling geographic information.

For the display and analysis of windscapes, GIS typically has multiple layers of data representing subsurface, surface, and lower atmospheric conditions. These datasets are obtained both from direct sampling and measurement as well as remote-sensing techniques. Such datasets would commonly include bedrock, soil, topography, vegetation and wildlife, wind conditions, human land use and ownership, transportation and utilities, protected or restricted areas, and related factors. Of these, three are fundamentally important for initial evaluation of wind-energy potential.

- Topography—Elevation of the land or shallow sea floor is the basic framework upon which wind turbines are erected and other natural phenomena and human land use take place. Elevation generally reflects underlying geologic structure and tectonic processes that have shaped the surface during past millennia, and the configuration of the land surface influences micro-climatic conditions and the resulting wind. Digital elevation or terrain models are GIS datasets that portray the land surface devoid of any other features (Fig. 2.4). A digital elevation model (DEM) is often utilized as the base map on which other features are depicted.
- Wind—The essential ingredient is wind, of course, specifically mean annual wind speed, as a measure of the energy potential for a site. Meteorological stations traditionally collect weather data near the ground surface where people live and work. Furthermore, many weather stations are located at civilian or military airports and may not be representative of surrounding areas. Such wind observations are seldom adequate for evaluating the energy potential of a proposed wind-farm site. Wind speed and consistency increase significantly with height above the ground (Gipe 2009), so it is important to have wind data from the turbine hub height, typically 80–120 m above the ground. For this purpose, a weather tower is erected and data are collected for a year or more before deciding to proceed with development of a wind farm. Meteorological measurements continue thereafter to evaluate performance of the turbines (Fig. 2.5).
- Electrical lines—Transmission refers to large, high-voltage lines that typically carry electricity from generating stations over long distances to major population or industrial centers. Distribution lines, on the other hand, are lower voltage for

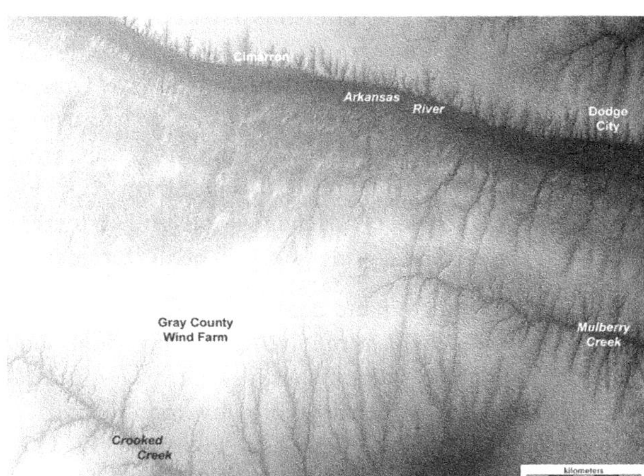

Figure 2.4 Shaded-relief digital elevation model for the Gray County Wind Farm vicinity, southwestern Kansas. Elevations range from less than 750 m (darkest gray) to above 850 m (white). The wind farm is located in the High Plains on the drainage divide between the Arkansas River and Crooked Creek at elevations 850–860 m. USGS/Public Domain.

Figure 2.5 Permanent meteorological tower in the Flat Ridge Wind Farm on the High Plains in south-central Kansas.

delivering electricity to local usage and individual customers. In general, alternating current (AC) transmission lines are 230 kV (kilovolts) or greater, and distribution lines are mostly 34.5 kV or less in the United States (Brown and Sedona 2004), although standard voltage levels vary from urban to rural and from country to country. Access to nearby transmission lines is crucial for developing wind energy, as such lines are quite costly to construct or upgrade (Fig. 2.6). In nearly all cases, wind farms also have dedicated electrical substations to connect with the regional grid system (Fig. 2.7). Thus, a GIS data layer for the electrical grid system is the third critical component.

These primary factors are combined with other basic geographic information for effective visual displays of wind-energy potential (Fig. 2.8). This is the basis for further evaluation of relevant issues that may affect decisions to develop wind energy. Beyond visual displays, GIS allows for selecting features, setting thresholds, and conducting statistical analyses to identify areas with the best combination of factors for producing wind energy as well as any potential negative impacts on the environment or human activities.

Figure 2.6 Spring Valley Wind Project in eastern Nevada feeds electricity into high-voltage transmission lines seen in the foreground. U.S. highway 50 crosses the valley in the left distance.

(A)

(B)

Figure 2.7 Electrical infrastructure for wind farms in Kansas. (A) substation for the Central Plains Project. (B) transformers and buried power lines in the Prairie Queen Project.

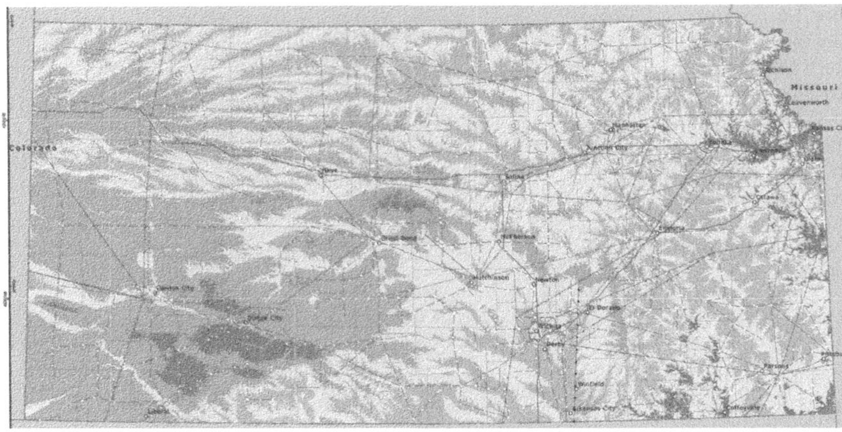

Figure 2.8 Combined GIS datasets for the wind resource of Kansas. Base map is shaded-relief elevation, color coded according to mean annual wind speed at 100 m height, ranging from pale gray (<7 m/s) to dark gray (>9 m/s). Transmission lines are marked according to their capacity along with major cities, highways, water bodies, and county boundaries. The state is approximately 400 miles (~600 km) across. KCC (2018)/ Kansas corporation commission/Public Domain.

A good example of this approach is The Nature Conservancy (TNC) evaluation of wind-energy impacts on wildlife in Kansas (Obermeyer et al. 2011). Improper siting and development of wind farms could impact wildlife through direct mortality as well as habitat loss and fragmentation. For a broader region, TNC's interactive GIS database includes multiple layers for 19 states in the central United States—Montana to Texas and New Mexico to Ohio (TNC 2022).

The map of key wildlife areas has more than a dozen layers, for instance, whooping crane stopover sites, bat roosts, intact natural habitats, and big game. Likewise, the map of potential engineering and land-use restrictions has 10 categories such as special use airspace, excessive slope, poor wind resource, and statutory setbacks. The combination of all these factors results in low-impact areas suitable for wind-energy development, which could lead to 1,000 gigawatts (GW) of renewable energy capacity (TNC 2022). Comprehensive as this analysis is for wind resource, engineering, land use, and wildlife, it still does not take into account available transmission lines or willing landowners.

The U.S. Geological Survey has undertaken a nationwide assessment of wind-energy impacts on wildlife, particularly for birds and bats, as well as habitat alterations (USGS Wildlife 2024). An essential ingredient for this assessment is the U.S. Wind Turbine Database (Hoen et al. 2018–2025), which was assembled from publicly available information and confirmed with visual identification in high-resolution remotely sensed imagery. The database is presented online in an

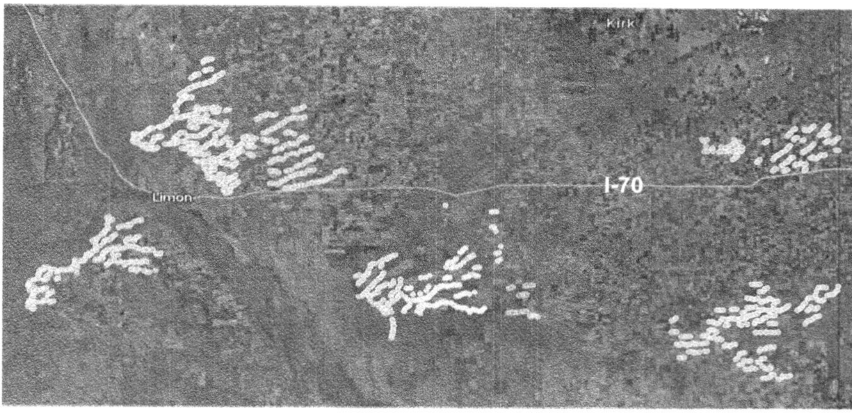

Figure 2.9 Wind farms along I-70 highway on the High Plains in east-central Colorado. Thirteen projects with 1,302 turbines have a combined capacity of 2,457 MW. Each dot represents a single turbine, color coded according to capacity. Map is approximately 170 km (105 miles) across; extracted and adapted from Hoen et al. (2018–2025)/USGS/Public Domain.

interactive GIS format that allows users to create and modify map displays as well as to download the wind-energy data. Individual turbines are located to within ±10 m and classified according to their technical specifications (Fig. 2.9).

Having emphasized the advantage of GIS for windscapes, it is wise to note some cautions. All maps are estimations, generalizations, and interpretations of true geographic conditions; no map could depict all physical, biological, and cultural features for even the smallest area. A map may display only a few selected features, which are portrayed usually in highly symbolic styles according to some kind of classification scheme. Furthermore, all maps and GIS datasets are products of human endeavor, which may lead to unwitting errors, misrepresentation, bias, or outright fraud. In spite of these limitations, maps and now GIS have proven to be remarkably adaptable and are necessary for display and evaluation of windscapes.

2.5 Economic, Social Science, and Environmental Methods and Techniques

Windscapes for producing energy both on land as well as offshore are often located in proximity to existing human populations where the electricity would be consumed. In many cases, wind turbines occupy landscapes or seascapes that already have long histories of human usage and modification (Fig. 2.10). In other cases, turbines may introduce negative impacts for natural environments,

Figure 2.10 Ramme Dige is a well-known archaeological site in western Denmark. It includes a cluster of burial mounds dating from the Stone Age and Bronze Age, one of which is visible on right side (<), as well as a wall-ditch structure (*) built during the Iron Age. Ramme wind farm is visible in the background (see sect. 6.2.3). Kite aerial photograph (2005).

particularly wildlife populations, human health, or aesthetic aspects. The economic impact, social acceptance, and value placed on energy projects in general, and wind energy in particular, may be assessed prior to initial development to ensure the viability of the project and the wider public's support for the same.

Quantifying and placing a value on wind-energy projects provide decision makers the tools with which to assess the broader economic impacts and the effectiveness of policies and programs. They also provide direction for future investments. Valuations allow for all costs and benefits to be tabulated to reflect the direct, indirect, and multiplier effects of any given wind-energy project. As Blanco (2009, p. 1373) outlined, wind-energy costs may include capital investment costs for turbines, project site construction, infrastructure, and grid connections. These may add up to 80% of the total costs of the project. Additional costs would include ongoing maintenance and operation costs plus rents, taxes, and insurance. As well, the cost of policy mechanisms and the subsidies (discussed in chap. 4) that support wind-power production must also be accounted for in such assessments.

The direct economic benefits of wind-energy projects are reflected in the sales of turbine parts, manufacturing, assembly and maintenance workers' earnings, while indirect benefits are accrued through manufacturing and employment in affiliated supplier sectors. For instance, chap. 4 describes the indirect impact of wind-energy policy on the steel industry in Brazil, which supplied turbine towers for projects across the country. Multiplier effects are garnered through ancillary spending and consumption that occurs across the economy through income generated from the target sector or project.

Renewable energy programs may have substantial economic benefits reflected in job growth, economic productivity, and income earnings. Numerous economic models are available to help quantify the influence of wind-energy policies and projects considering everything from the local to macro-scale impacts. Some of the more-often-used modeling methods include input–output techniques or other econometric approaches.

Input–output methods may help estimate the ripple effects across an economy of changes in one economic activity. For example, an economic activity may include a new wind-power project or a new government incentive for wind-energy production. Input–output methods require detailed data on project construction, maintenance and labor costs, as well as other project financing parameters. The models are, however, static representations and provide one snapshot in time, though they may be built with relative ease. Input–output models come with a number of other limitations including their assumptions of fixed prices and fixed production patterns and homogeneity across industry sectors. If unaccounted for, these may lead to erroneous results (Bess and Ambargis 2011). Even so, ease of use has allowed for input–output models to be employed when quick assessments of an initial industry change are required.

The National Renewable Energy Laboratory's (NREL) Wind Powering America program released the Jobs and Economic Development Impact (JEDI) for land-based wind model in 2009, and an offshore wind model was introduced in 2021 (NREL 2024). These models use input–output tools to estimate the broader economic impacts of constructing and operating a wind-power generation facility at local and state scales. JEDI reports employment, earnings, and output generated through direct project-related development and labor impacts, as well as indirect industry and supply-chain impacts and other ripple or induced effects.

Econometric models are far more sophisticated, detailed, and built to account for changes over time. They consider historical and cross-sectoral data interactions to help predict future impacts. These may also include statistical methods to evaluate the significance and robustness of predictions. For instance, the PANTA RHEI macro-economic simulation forecasting model, which also incorporates environmental conditions, has been used to evaluate the economic impact of renewable energy on the labor market (Lutz 2011; Lehr et al. 2012). These highly detailed econometric modeling techniques are employed to consider the macro-economic impacts of wind-energy policies and programs.

In addition to economic accounting, other social-science approaches have been used to quantify those costs and benefits that do not have fixed monetary values or remain elusive to traditional cost-accounting methods. The Social Life Cycle Assessment (S-LCA) method used by Lehmann et al. (2022) and others offers novel approaches to evaluating the broader implications of wind-energy projects by prioritizing social impact categories and stakeholders. There is a crucial role of embedding such social impact assessments to empower local communities participating in community energy projects (Muthoora and Fischer 2019). This integration is essential for the broader acceptance and sustainability of wind-energy projects. Such an approach was employed in a study on a wind-power plant in southern Sardinia. It analyzed sustainability indicators related to plant site characteristics, technical features, stakeholder opinions, environmental and social impacts, and expected economic benefits, and offered a comprehensive understanding of the project's impacts (Fois et al. 2022).

Integrative frameworks and participatory approaches are especially relevant to assessing ecosystem and other social services provided by natural resources. For instance, respondents may be questioned on their economic commitment to setting aside forested land from development activity to reap the ecosystem benefits provided by such an act. Similarly, wind-power projects and policy have been assessed using contingent valuation (CV) methods, which can estimate the general public's acceptance and willingness to pay for renewable energy. These insights may be informative in assessing the value placed on the development of wind energy or the feasibility of particular projects.

Surveys designed to gauge contingent value must have a clearly defined resource scenario around which a set of questions are posed. For example, a wind-energy project that assesses the public's willingness to pay a surcharge on monthly energy bills in support of renewable energy must precisely outline the benefits garnered through such a project. The survey must also provide a realistic scenario with payments that align as close as possible to actual payments. Such approaches have been shown to improve survey reliability and validity (Byrnes et al. 1999; Longo et al. 2008). Moreover, sampling frameworks using random, systematic, stratified, or cluster sampling techniques may provide statistically robust ways to collect such data from a smaller subset of a larger population.

Studies indicate that a respondent's social and environmental attitudes, demographic and socio-economic parameters, and political leanings could all influence survey responses (Menegaki 2008). Moreover, many studies have found that responses do not always tally with actual actions (Byrnes et al. 1999). Respondents are often universally supportive of renewable energy, yet when it comes to actually paying associated surcharges, they may be less inclined to commit. These are shortcomings with CV methods. Even so, once responses to the questions are processed, they could be used to at least estimate support for the resource being assessed.

Beyond willingness to pay assessments, surveys can also be used to gauge the social acceptance of wind-energy projects. The social acceptance literature is particularly large in the case of wind energy, and studies find that while general acceptance of renewables is high, there is divergence as one focuses on communities and locales directly affected by wind projects (Fast 2013). Such divergence, however, is far from universal. While some European studies indicate higher rates of acceptance, studies from elsewhere still suggest a not-in-my-backyard (NIMBY) approach (Jones and Eiser 2010; Musall and Kuik 2011). A region's history with wind development would play an influential role in wider acceptance. Meanwhile, places, and windscapes in particular, have cultural and personal significance, which may also lead to the disapproval of projects elsewhere.

In addition to CV or acceptance surveys, in-depth interviews and oral histories with stakeholders and affected citizens may often produce detailed and valuable information on windscape issues under scrutiny. Moreover, matters are more complicated in the developing world when wind-energy projects are proposed on land that may be held as a common property resource. In such cases, in-depth interviews could be particularly valuable to minimize the negative impacts of wind-project exclusion zones that may impinge on local communities.

India, given its lengthy coastline, is looking at offshore wind-energy production as a promising source to help increase its energy security. While offshore wind production is relatively new to the country, a 30-GW target by 2030 would mean rapid expansion, which could have significant impacts for local coastal communities who depend on fishing for their sustenance and livelihoods (Sarangi 2022). In such instances, data from interviews and other social-science assessment methods could provide decision makers with important insights with which to assess policy and management alternatives or the viability of specific projects.

Over the past decade, quantifying the impacts of wind-energy projects on ecosystems and the services they provide have become commonplace. Environmental Impact Assessments (EIA) are often required before large-scale projects can be implemented. Emerging from many of these studies is the need for enhanced monitoring, mitigation strategies, and integrative-planning approaches to address the ecological concerns associated with wind-energy projects. For instance, Guşatu et al. (2021) assessed the potential impacts of offshore wind farms on the marine ecosystem in the North Sea basin, in light of the significant expansion expected to meet renewable energy goals. Their geospatial analysis highlights the importance of considering cumulative and long-term environmental effects in the planning and development of offshore wind-energy projects.

In another study, Rostin et al. (2013) documented the impacts to the marine ecosystem of wind-energy projects across the Baltic Sea that must be taken into consideration both during the construction and the operational phases. Local and global assessments of environmental considerations may be helpful to provide a holistic picture of the impacts, both positive and negative of the wide-scale implementation of wind energy (Wang et al. 2015; Wawrzyczek et al. 2018; Wu et al. 2024). Effective environmental governance at the country and international scale, comprehensive impact assessments, and stakeholder engagement are crucial elements in reconciling the objectives of renewable energy expansion with social and ecosystem considerations.

2.6 Summary

The windscape represents the combination of natural conditions and human exploitation of a site for wind energy. Means of investigating and analyzing the windscape vary widely for particular disciplines and purposes. Geospatial analysis, also known as geographic information systems or science (GIS), is a powerful means to integrate physical, biological, and cultural aspects of the windscape. Datasets collected on the ground by direct measurement and surveying are combined with aerial observations, satellite imagery, and other forms of remote sensing. Scale and resolution are key aspects to geographic information, ranging from micro- to meso- to macro-landscapes.

In the assessment of windscapes, three factors are fundamentally important for initial evaluation of wind-energy potential—topography, wind, and transmission lines. These primary factors are combined with other geographic information and form the basis for further evaluation of relevant issues that may affect decisions to develop wind energy. Windscapes are often located in proximity to existing human populations, and wind turbines occupy landscapes or seascapes that already have long histories of human usage and modification. Recognizing and documenting the cultural and socioeconomic values of windscapes are, thus, vital to their suitable management.

Economic valuation methods including input–output or other econometric models assess the broader impacts of particular wind-energy projects. CV methods including willingness to pay or social acceptance data may be measured through survey and interview data. Such methods can be used to assess the use, value, and impacts of windscapes. EIA help highlight the complexity of immediate and long-term ecosystem impacts associated with wind-energy projects.

References

Aber, J.S., Aber, S.W. and Pavri, F. 2015. Windscapes: A global perspective on wind power. Multi-Science Publ., UK, 245 p.

Aber, J.S., Marzolff, I., Ries, J.B. and Aber, S.E.W. 2019. Small-format aerial photography and UAS imagery: Principles, techniques and geoscience applications, 2nd ed. Elsevier, Amsterdam, Netherlands, 382 p.

Aber, J.S., Pavri, F. and Aber, S.W. 2012. Wetland environments: A global perspective. Wiley-Blackwell, 421 p.

Aber, J.S., Zupancic, J. and Aber, S.W. 2003. Applications of kite aerial photography: Golf course management. *Kansas Academy of Science, Transactions* 106, p. 211–214.

Aber, S.E.W. and Aber, J.W. 2017. Map librarianship: A guide to geoliteracy, map, and GIS resources and services. Chandos Publ., Elsevier, Cambridge, Massachusetts, 278 p.

Avery, T.E. and Berlin, G.L. 1992. Fundamentals of remote sensing and airphoto interpretation, 5th ed. Macmillan, New York, 474 p.

Bess, R. and Ambargis, Z. 2011. Input-output models for impact analysis: suggestions for practitioners using RIMS II multipliers. 50th Southern Regional Science Association Conference, March 23–27, 2011, New Orleans, Louisiana. https://www.bea.gov/system/files/papers/WP2012-3.pdf

Blanco, M. 2009. The economics of wind energy. *Renewable and Sustainable Energy Reviews* 13, p. 1372–1382. https://doi.org/10.1016/j.rser.2008.09.004

Brittan, G.G. Jr. 2002. A view from Lake Como. In: M. Pasqualetti, P. Gipe and R. Righter (eds.), Wind power in view: Energy landscapes in a crowded world. Academic Press, Elsevier, p. 215–221.

Brown, M.H. and Sedano, R.P. 2004. Electricity transmission: A primer. National Council on Electricity Policy. https://www.energy.gov/oe/articles/electricity-transmission-primer

Burrough, P.A. 1986. Principles of geographical information systems for land resources assessment (reprinted with corrections). Monographs on soil and resources survey 12. Oxford University Press, Oxford, United Kingdom, 194 p.

Byrnes, B., Jones, C. and Goodman, S. 1999. Contingent valuation and real economic commitments: Evidence from electric utility green pricing programmes. *Journal of Environmental Planning and Management* 42(2), p. 149–166. https://doi.org/10.1080/09640569911190

Campbell, J.B., Wynne, R.H. and Thomas, V.A. 2022. Introduction to remote sensing, 6th ed. Guilford Press, New York, 634 p.

Fast, S. 2013. Social acceptance of renewable energy: Trends, concepts, and geographies. *Geography Compass* 7(12), p. 853–866. https://doi.org/10.1111/GEC3.12086

Fois, V., Lecis, L. and Cocco, D. 2022. Social impact assessment of wind power generation. An innovative method for decision making processes. *Journal of Physics: Conference Series* 2385, 012004. https://doi.org/10.1088/1742-6596/2385/1/012004

Gipe, P. 2009. Wind energy basics: A guide to home- and community-scale wind energy systems, 2nd ed. Chelsea Green Publishing, White River Junction, Vermont, 166 p.

Guşatu, L.F., Menegon, S., Depellegrin, D., Zuidema, C., Faaij, A. and Yamu, C. 2021. Spatial and temporal analysis of cumulative environmental effects of offshore wind farms in the North Sea basin. *Scientific Reports* 11, p. 10125. https://doi.org/10.1038/s41598-021-89537-1

Hall, R.C. 1997. Post war strategic reconnaissance and the genesis of project Corona. In: R.A. McDonald (ed.), Corona: Between the sun & the earth: The first NRO reconnaissance eye in space. American Society of Photogrammetry and Remote Sensing, p. 25–58.

Hoen, B.D., Diffendorfer, J.E., Rand, J.T., Kramer, L.A., Garrity, C.P. and Hunt, H.E. 2018. United States Wind Turbine Database V8.0 (February 25, 2025). U.S. Geological Survey, American Clean Power Association, and Lawrence Berkeley

National Laboratory data release. https://energy.usgs.gov/uswtdb/ or https://doi.org/10.5066/F7TX3DN0

Jensen, J.R. 2007. Remote sensing of the environment: An earth resource perspective, 2nd ed. Prentice Hall Series in Geographic Information Science, Upper Saddle River, New Jersey, 592 p.

Jones, C. and Eiser, J. 2010. Understanding 'local' opposition to wind development in the UK: How big is a backyard? *Energy Policy* 38(6), p. 3106–3117. https://doi.org/10.1016/j.enpol.2010.01.051

KCC 2018. Wind resource of Kansas: Mean annual wind speed at 100 meters. Kansas Corporation Commission, Kansas wind maps. https://www.kcc.ks.gov/images/PDFs/kansas-energy-office/Wind_WindFarmswithWindSpeeds.pdf

Lauer, D.T., Morain, S.A. and Salomonson, V.V. 1997. The Landsat program: Its origins, evolution, and impacts. *Photogrammetric Engineering & Remote Sensing* 63, p. 831–838. https://www.asprs.org/wp-content/uploads/pers/1997journal/jul/1997_jul_831-838.pdf

Leachtenauer, J.C., Daniel, K. and Vogl, T.P. 1997. Digitizing corona imagery: Quality vs. cost. In: R.A. McDonald (ed.), Corona: Between the sun & the earth, the first NRO reconnaissance eye in space. American Society Photogrammetry and Remote Sensing, p. 189–203. https://www.asprs.org/wp-content/uploads/pers/1998journal/jan/1998_jan_29-34.pdf

Lehmann, J., Bouillass, G., Fofack-Garcia, R. and Pérez-López, P. 2022. Towards social life cycle assessment of energy systems: A case study on offshore wind farms from companies' perspective. *E3S Web of Conferences* 349, p. 12002. https://doi.org/10.1051/e3sconf/202234912002

Lehr, U., Lutz, C. and Edler, D. 2012. Green jobs? Economic impacts of renewable energy in Germany. *Energy Policy* 47, p. 358–364. https://doi.org/10.1016/j.enpol.2012.04.076

Longo, A., Markandya, A. and Petrucci, M. 2008. The internalization of externalities in the production of electricity: Willingness to pay for the attributes of a policy for renewable energy. *Ecological Economics* 67, p. 140–152. https://doi.org/10.1016/j.ecolecon.2007.12.006

Lutz, C. 2011. Energy scenarios for Germany – Simulations with the model PANTA RHEI. In: D. Mullins, J. Viljoen and H. Leeuwner (eds.), Interindustry based analysis of macroeconomic forecasting. Conningarth Consulting Economists, Pretoria, South Africa, p. 203–224. http://inforumweb.inforumecon.com/papers/conferences/2011/germany_lutz.pdf

Menegaki, A. 2008. Valuation for renewable energy: A comparative review. *Renewable and Sustainable Energy Reviews* 12(9), p. 2422–2437. https://doi.org/10.1016/j.rser.2007.06.003

Merriam, D.F. 1996. Kansas 19th century geologic maps. *Kansas Academy of Science, Transactions* 99, p. 95–114. https://www.jstor.org/stable/3627983

Musall, F. and Kuik, O. 2011. Local acceptance of renewable energy: A case study from southeast Germany. *Energy Policy* 39(6), p. 3252–3260. https://doi.org/10.1016/j.enpol.2011.03.017

Muthoora, T. and Fischer, T.B. 2019. Power and perception – From paradigms of specialist disciplines and opinions of expert groups to an acceptance for the planning of onshore windfarms in England – Making a case for Social Impact Assessment (SIA). *Land Use Policy* 89, 104198. https://www.sciencedirect.com/science/article/pii/S0264837718319641?via%3Dihub

NREL. 2024. JEDI wind models. National Renewable Energy Laboratory. https://www.nrel.gov/analysis/jedi/wind.html

Obermeyer, B., Manes, R., Kiesecker, J., Fargione, J. and Sochi, K. 2011. Development by design: Mitigating wind development's impacts on wildlife in Kansas. *PLoS One* 6(10), e26698. https://doi.org/10.1371/journal.pone.0026698. https://journals.plos.org/plosone/article/file?type=printable&id=10.1371/journal.pone.0026698

Pavri, F. and Aber, J.S. 2004. Characterizing wetland landscapes: A spatiotemporal analysis of remotely sensed data at Cheyenne Bottoms, Kansas. *Physical Geography* 25(1), p. 86–104. https://doi.org/10.2747/0272-3646.25.1.86

Philipson, W.R. (ed.). 1997. Manual of photographic interpretation, 2nd ed. American Society for Photogrammetry and Remote Sensing, Bethesda, Maryland.

Rostin, L., Martin, G. and Herkül, K. 2013. Environmental concerns related to the construction of offshore wind parks: Baltic Sea case. *WIT Transactions on Ecology and the Environment* 169, p. 131–140. https://doi.org/10.2495/CP130121

Sarangi, G. K. 2022. Offshore wind energy as an emergent ocean infrastructure in India: Mapping of the social and environmental impacts. ADBI Working Paper 1307. Tokyo: Asian Development Bank Institute. https://www.adb.org/sites/default/files/publication/778186/adb-wp1307.pdf

Tatem, A.J., Goetz, S.J. and Hay, S.I. 2008. Fifty years of earth-observation satellites. *American Scientist* 96, p. 390–398. https://www.americanscientist.org/article/fifty-years-of-earth-observation-satellites

TNC 2022. Site renewables right: Accelerating a clean and green renewable energy buildout in the central United States. The Nature Conservancy. https://www.nature.org/content/dam/tnc/nature/en/documents/SRR_Methods_20220202_LR.pdf

USGS Landsat 2022. Landsat 9 image of San Francisco Bay, California. U.S. Geological Survey. https://www.usgs.gov/media/images/landsat-9-image-san-francisco-bay-california

USGS Wildlife 2024. Can wind turbines harm wildlife? U.S. Geological Survey. https://www.usgs.gov/faqs/can-wind-turbines-harm-wildlife

Wang, S., Wang, S. and Smith, P. 2015. Quantifying impacts of onshore wind farms on ecosystem services at local and global scales. *Renewable and Sustainable Energy Reviews* 52, p. 1424–1428. https://doi.org/10.1016/j.rser.2015.08.019

Wawrzyczek, J., Lindsay, R., Metzger, M. and Quétier, F. 2018. The ecosystem approach in ecological impact assessment: Lessons learned from windfarm developments on peatlands in Scotland. *Environmental Impact Assessment Review* 72, p. 157–165. https://doi.org/10.1016/j.eiar.2018.05.011

Wu, G., Deshmukh, R., Trainor, A., Uppal, A., Chowdhury, K.A., Baez, C., Martin, E., Higgins, J., Mileva, A. and Ndhlukula, K. 2024. Avoiding ecosystem and social impacts of hydropower, wind, and solar in Southern Africa's low-carbon electricity system. *Nature Communications* 15(1). https://doi.org/10.1038/s41467-024-45313-z

3

Converting Wind into Electricity

Small changes in wind speed have a profound effect on the power in the wind (Gipe 2009, p. 42).

3.1 Atmosphere and Wind

The atmosphere is the ocean of air that surrounds the Earth. It is composed primarily of nitrogen, oxygen, water vapor, argon, carbon dioxide, and many other trace elements and compounds in gaseous form. The atmosphere typically also has liquid and solid components, namely water droplets, ice crystals, and dust from diverse natural and human sources. The gas composition of the atmosphere is nearly constant around the world, but the liquid and solid components vary greatly by location and through time. Likewise, the physical properties of the atmosphere, namely its pressure and temperature, also experience large changes in time and space. Wind is the atmosphere in motion, and any material in motion possesses energy, specifically kinetic energy, which may be captured and converted into electricity.

3.1.1 Wind Basics

Wind is a consequence of pressure differences within the atmosphere. Put simply, wind flows from zones of high pressure toward low pressure. Air pressure, in turn, is largely a function of temperature differences in the atmosphere; warm air has a lower density and so rises creating low pressure, whereas cold air is denser and sinks leading to high pressure. Surface wind represents the transfer of air masses

Wind Energy Landscape: Principles, Techniques, History, First Edition. James S. Aber, Susan E. W. Aber, and Firooza Pavri.

Published 2026 by John Wiley & Sons, Inc.

from zones of cold sinking air toward zones of warm rising air. The greater the difference in temperature, the larger the pressure gradient, and the stronger wind that results.

Overall, the atmosphere is warmer at lower latitudes in the tropical zones near the Equator and colder at higher latitudes toward the poles (Fig. 3.1). This sets up zonal wind patterns that are consistent over broad regions of the globe (Fig. 3.2).

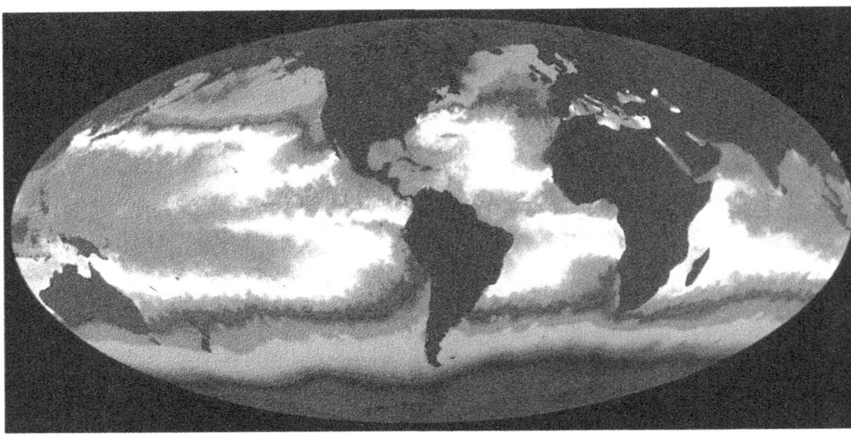

Figure 3.1 Global sea-surface temperature for the period July 1984. Measured by the Advanced Very High Resolution Radiometer (AVHRR) on board the NOAA-7 satellite. Adapted from NASA; obtained from Wikimedia Commons http://commons.wikimedia.org/

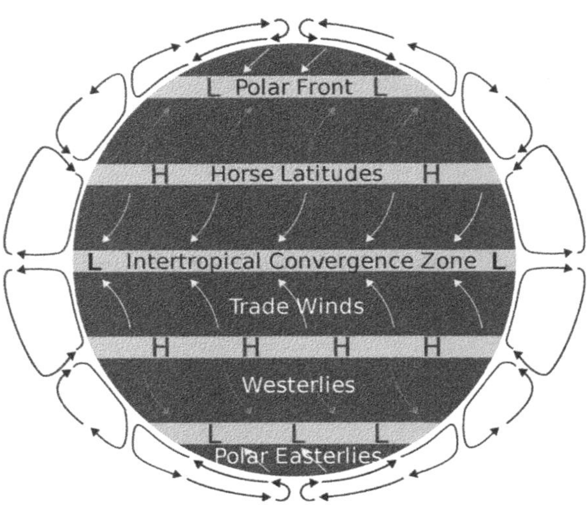

Figure 3.2 Schematic diagram of global atmospheric circulation. H: high pressure, L: low pressure. Adapted from http://commons.wikimedia.org/

- Trade winds—blowing from the northeast in northern hemisphere and southeast in southern hemisphere toward the low-pressure Intertropical Convergence Zone.
- Horse latitudes—weak and variable wind. Zones of high pressure approximately 30° north and south latitudes; the main desert belts.
- Westerlies—blowing from the northwest or southwest depending on the season in the middle latitudes, 30–60° north and south.
- Polar easterlies—blowing from the northeast in northern hemisphere and southeast in southern hemisphere toward the low-pressure Polar Front.

These zonal wind patterns tend to be quite stable over oceanic regions, because sea-surface temperature varies only slightly throughout the year. Land-surface temperatures, on the other hand, change considerably through the seasons, which gives rise to more variable wind conditions, especially over the interiors of larger continents. The placements of continents, large islands, mountain systems, broad plateaus and plains, ice sheets, inland seas, large lakes, and other major geographic features influence further complications in local and regional wind patterns.

3.1.2 Diurnal and Seasonal Winds

Wind is one means by which thermal energy from the sun is transported from low latitudes and altitudes to high latitudes and altitudes. Ocean currents are the other primary means for moving heat around the globe. Over land regions, near-surface air temperatures fluctuate from day to night with heating and cooling of the ground. This gives rise to diurnal (day–night) wind patterns. For example, in continental interiors, such as the Great Plains of North America, wind is often nearly calm at sunrise and early morning. As the ground warms during the day, wind speed increases and typically reaches maximum velocity in the afternoon. As evening approaches, wind begins to lessen and may dissipate during the night. The peak interval for potential wind-energy, thus, is during the hottest part of the day when electricity for air conditioning is most in demand during summer months.

Coastal regions may display quite different diurnal winds, however. During the hottest part of the day, the land area is relatively warm compared with the adjacent water body. This causes warm, low-density air to rise over the land, and cooler air flows inland as a refreshing sea breeze. During night, the process reverses, and the land becomes cold relative to the water body; offshore wind carries the cooler land air out to sea. This effect is especially pronounced in coastal deserts or mountains that have large swings in diurnal land temperature. Passing storm systems may interrupt the normal diurnal pattern, however, both inland and along coasts.

A strange tale has gained traction during the past quarter century, namely that wind turbines generate more electricity at night. This is simply wrong. On a global basis, peak wind speed and electricity generation are achieved on a long-term average during daylight hours (DWIA 2003; Aber et al. 2015).

Myth of Diurnal Wind

According to Graetz (2011, p. 191), *wind turbines . . . tend to generate more electricity at night*. This is an oft-repeated claim that is accepted uncritically, year after year even by reputable sources, and is considered a serious limitation for wind energy (e.g. Dizikes 2010; Bernardos and Fulda 2020). While this may be true in some circumstances, it is simply wrong nearly everywhere. In fact, at *most locations around the globe it is more windy during the daytime than at night* (DWIA 2003). Simply put, diurnal wind speed generally follows the sun and daily temperature fluctuations.

Afternoon peak in average daily wind speed is, in fact, the norm in most locales most of the time around the world. In Luxembourg, for instance, Massen et al. (2007) documented wind conditions in connection with a long-term study of atmospheric CO_2. They found a strongly diurnal pattern in which minimum average wind speed happens overnight; whereas, maximum average wind speed takes place in the afternoon.

Many examples may be cited around the world (Aber et al. 2015). The island of Puerto Rico is located in the trade-wind belt of the Caribbean Sea. It experiences maximum daily average wind speed during the afternoon hours. Long-term diurnal wind pattern is similar for Ireland. For tropical Sri Lanka in the Indian Ocean, average daily wind speed peaks in the late afternoon. This is true likewise in the high-altitude desert of the San Luis Valley in south-central Colorado of the United States.

A similar diurnal pattern for early afternoon maximum average daily wind speed is documented for Waterloo, Canada, in all seasons of the year (Li and Li 2005). Daily average wind speed peaks in the mid-afternoon at Florissant, Colorado, at 2,635 m altitude in the Rocky Mountains (CSU 2013). For Beijing, China, average diurnal wind speed reaches a maximum in the afternoon and is minimal overnight (Yan et al. 2021). In tropical, rural, southern India, peak diurnal wind speed happens at mid-day during the summer (May–September), but shifts to late afternoon during other months (October–April) according to Reddy et al. (2014). Finally the case may be cited of the Atacama Large Millimeter/submillimeter Array at 5,000 m altitude in Chile, where average daily wind speed peaks in the early afternoon (ALMA 2025).

As these many examples drawn from around the world demonstrate, long-term, average, diurnal wind usually reaches peak velocity during the afternoon hours for nearly all locations. Passing storm systems or other atmospheric disturbances may, of course, upset this pattern for short time periods. Nonetheless, wind-energy production is based on long-term averages, not short-term deviations. Thus, the blanket claim that wind energy is produced mainly at night must be regarded as a misguided myth or perhaps a disingenuous attempt to discredit wind energy.

Many parts of the world have well-known winds that blow on a seasonal basis, for example, the Harmattan in northern and western Africa, the Cierzo in northern Spain, and the Mistral in the western Mediterranean region. All of these take place during the dry winter, but in summer these regions are dominated by calm or thermally induced local winds (Aber et al. 2019). Similar seasonal winds are encountered in many parts of the world, for example the Santa Ana, a dry desert wind from the east or northeast in southern California usually during cooler months of the year.

3.1.3 Air Density and Altitude

A majority of humanity lives at relatively low elevations below 300 m, and nearly 90% live below 1,000 m altitude (Cohen and Small 1998). Thus, it is no surprise that most wind energy has been developed at low elevations. Nonetheless, some wind farms are situated at relatively high altitudes where air density is considerably less than at sea level. The force that affects an airfoil, such as an airplane wing, kite, or turbine blade, is determined by relative wind speed and air density acting on the lifting surface. At high altitude, lower air density means that fewer and lighter molecules (per air volume) flow over the airfoil at a given wind speed.

Air density is governed by three factors—pressure, temperature, and humidity. The combination of these factors determines potential lifting power of a particular airfoil for a given wind speed. The standard atmosphere is an ideal model of atmospheric conditions (Table 3.1). These values reveal that density decrease is fairly slight up to about 1,000 m above sea level. At higher altitudes, however, air density declines significantly, especially at higher temperature. For instance, Tehachapi Pass, California, at ~1,500 m altitude, on a hot day (35 °C), air density is only about 80% of that on the Pacific Coast at 20 °C (Gipe 2009). Wind farms in the western United States are now routinely built at and above 1,800 m altitude (Fig. 3.3). Turbine makers have to take the thin air at high altitude into account for turbine and blade design as well as estimating likely energy output.

Table 3.1 Standard atmospheric conditions for temperature, pressure, density, and density percentage according to altitude up to 3,000 m.

Altitude		Temp. (C)	Pressure (hPa)	Density (kg/m^3)	Percent density
Sea level (m)	Sea level (feet)	15.0	1,013	1.2	100
1,000	3,280	8.5	900	1.1	92
2,000	6,560	2.0	800	1.0	83
3,000	9,840	−4.5	700	0.91	76

Aber et al. (2015).

Figure 3.3 GE Wind turbines and transmission line in the Pioneer Wind Park near Glenrock in central Wyoming. The wind farm occupies a high plateau of granite and other crystalline bedrock on the northern flank of the Laramie Mountain range. Elevation is 2,000 to 2,100 m, which rises some 500 m above the North Platte River valley at Glenrock.

3.1.4 Near-Surface Wind

Near the surface of the Earth, wind interacts with the land and water boundary to create turbulent air flow over and around ground objects. Landscape topography, vegetation cover, human structures, and land use all influence near-surface wind conditions. Turbulent flow dissipates energy and slows the wind near the ground. Such effects are most obvious in large urban settings, forests, and terrain with substantial topographic obstacles (Fig. 3.4). For an individual obstacle, such as a building or tree, a zone of ground turbulence extends downwind a distance 20 times the height of the object and vertically two times the height of the object (Fig. 3.5; Gipe 2009). The variation of wind speed (V) with height (H) above the ground is demonstrated by an empirical power law (Musgrove 2010):

$$V / V_{10} = \left(H / 10\right)^{\alpha}$$

where V_{10} is the reference wind speed at 10 m above the ground. The influence of ground cover is expressed by the roughness exponent (α), which varies from less than 0.1 for smooth surfaces to >0.4 above forest canopies (Table 3.2). For these reasons, wind farms are typically situated on hill tops or drainage divides in open terrain with low vegetation and few human structures—desert, cropland, or

Figure 3.4 Tall buildings, towers, and numerous trees reduce average near-surface wind speed over large urban areas compared with their rural surroundings. Urban surface wind is also typically more gusty and variable in direction. Helium-blimp aerial photograph; mid-town Kansas City, Missouri. The television tower stands 318 m tall and dates from 1955.

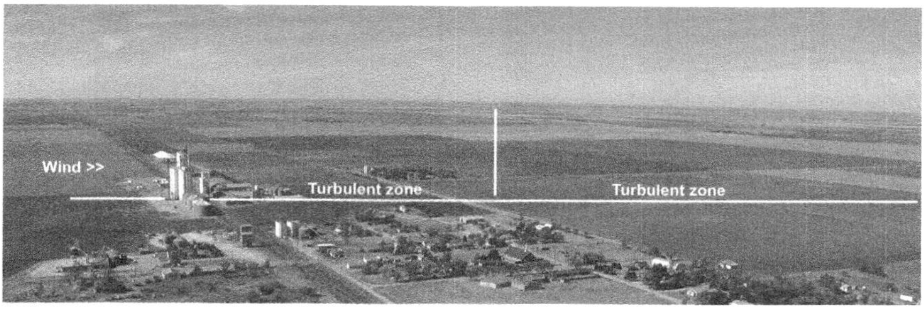

Figure 3.5 Wind blowing from left to right around the grain elevator standing on the left side. A zone of turbulence extends two times the elevator height (*H*) upwind and vertically, and 20 times the elevator height downwind, as indicated schematically by the white lines. Kite aerial photograph; city of Marienthal on the High Plains of western Kansas.

grassland (Fig. 3.6). In other cases, wind gaps in mountain ranges, between plateaus, or through hilly terrain may serve to focus near-surface wind flow as demonstrated by the main California wind-energy complexes (see sect. 1.4.2).

Table 3.2 Terrain cover features and typical roughness exponent (α) at a height of 10 m.

Terrain cover	Roughness exponent
Snow on flat ground	0.07
Calm sea	0.09
Onshore coastal wind	0.11
Snow-covered crop stubble	0.12
Mowed grass	0.14
Shortgrass prairie	0.16
Crops, tallgrass prairie	0.19
Hedges	0.21
Scattered trees & hedges	0.24
Trees, hedges, buildings	0.29
Suburban landscape	0.31
Woodland, forest	0.43

Aber et al. (2015).

Figure 3.6 Vestas V90-1.8 MW turbines of the Caney River Wind Project are situated in tallgrass prairie on a prominent cuesta at the eastern edge of the Flint Hills in south-central Kansas. The ridge tops stand more than 75 m above adjacent lower terrain. Kite aerial photograph with J. Schubert.

Drainage Divides

In any land-based setting, drainage divides are typically the highest topographic features present. Such divides may stand 10s to 100s of meters above adjacent river valleys and lowlands. On a regional basis, divides separate major drainage basins, and the divides may extend for 100s of km up and down stream. Given their topographic prominence and exposed positions, drainage divides typically have the best local prospects for wind energy.

Drainage Divides (Continued)

This is particularly true where resistant bedrock ridges, glacial moraines, or other geologic landforms enhance the topographic relief of drainage divides (e.g. Buffalo Ridge, see sect. 1.4.3). In eastern Montana, United States, for instance, a major drainage divide separates the Missouri River and Yellowstone River basins. The Clearwater complex of multiple wind farms occupies this divide at elevations exceeding 900 m, more than 160 m above the Yellowstone River valley to the south.

The most prominent drainage divide in the central Great Plains of the United States separates the Missouri River and Arkansas River basins. This divide begins in eastern Colorado, crosses Kansas, and extends into southwestern Missouri. A total of seven wind farms are situated along this divide in Kansas, and two more wind farms occupy the divide in Missouri, a total distance of approximately 650 km. These wind farms date from 2009 to 2021 and include a total of 663 turbines with a generating capacity of 1,687 MW (Hoen et al. 2018–2025).

As a specific example, the Jayhawk wind farm in southeastern Kansas has 70 GE Wind turbines with a combined generating capacity of 197 MW. The western portion of the wind farm is situated on a prominent escarpment formed by the Bethany Falls Limestone at elevations reaching 335 m (NGMD 2025). The eastern part, in contrast, is about 30 m lower in elevation.

GE Wind turbines in the Clearwater project on the Missouri-Yellowstone drainage divide approximately 55 km north of Miles City, Montana.

(*Continued*)

Drainage Divides (Continued)

This vicinity is largely agricultural with a mix of row crops and pasture. Installation of the turbines and access roads has taken some farm land and requires increased maneuvering of farm implements in crop fields. But the minimal impact on agriculture is a small price to pay compared with the income land owners derive from the wind turbines.

Drainage basins and Missouri–Arkansas divide (dots) across Kansas into southwestern Missouri. Nine wind farms (*) are situated along this major divide. Drainage basins based on U.S. Geological Survey.

Western, higher part of Jayhawk wind farm near Porterville, Kansas. On the drainage divide between the Missouri River and Arkansas River basins. Total height of GE Wind turbines with a blade in upright position is 177.4 m (582 feet; Hoen et al. 2018–2025). Autumn kite aerial photograph (2022).

Figure 3.7 Adjacent active and rotating turbines face in different directions at nearly right angles. Vestas V120 turbines in the Busch Ranch II wind farm on the High Plains near Walsenburg, Colorado.

Local topographic features and ephemeral weather conditions could affect near-surface wind direction and speed, such that individual turbines respond in ways that appear peculiar. Within a seemingly uniform array of turbines, some may shut down while others continue to spin at normal speed, and nearby turbines may face in quite different directions (Fig. 3.7). This reflects the turbulent nature of wind and the autonomous operation of each turbine in response to wind measured by sensors on the nacelle.

3.1.5 Wind Droughts and Climate Change

The term wind drought was coined in connection with a prolonged interval of low wind speed in the United States during the first three months of 2015, and this substantially reduced electric-power generation of wind farms. Similar calm periods on the high seas have been known to sailors for centuries. Nonetheless, wind droughts are yet another aspect to take into consideration about the viability of wind energy.

In general, it is well known that average wind speed over North America is related to climatic conditions in the Pacific region. The 2015 episode of low wind has been attributed to the North Pacific Mode and more specifically to high sea-surface temperatures (Lledó et al. 2018). A similar wind drought took place in the United Kingdom, which has most of its wind turbines deployed offshore in the North Sea. During summer and early autumn of 2021, the United Kingdom suffered a wind drought in which production of wind energy declined by nearly one-third of normal (Bloomfield 2021).

It should be noted that power generated by a turbine is related to the cube of the wind speed; thus, small changes in average wind speed have large consequences for generating electricity (see below). Projected global warming may lead to long-term reduction in wind speed in some areas, but may cause increases in other places, according to some climate models (Bloomfield 2021). In general, much of the western and eastern United States may experience decreased average wind speeds. On the other hand, the central United States could develop increased wind in some seasons, particularly for the southern Great Plains (Chen 2020).

Indeed, a new scientific discipline has emerged, known as energy meteorology, in which the atmosphere and climate are viewed as resources, particularly for wind and solar energy (Olsson 1994). As our understanding of climate improves, especially wind variability and droughts, so would our decisions about wind-energy deployment and operation around the world.

3.2 Wind Power and Energy

In common practice, the terms energy and power are often used interchangeably. But strictly speaking, power is the rate at which energy is utilized. Horsepower (hp) is one common example of power measurement. James Watt and Matthew Boulton greatly enhanced steam-engine design in the late 1700s, and Watt introduced hp as a way to describe the power of steam engines. The unit of power for electricity, is the watt (W), named after James Watt (Musgrove 2010). One metric hp = 735.5 W.

Most people are still familiar with conventional incandescent bulbs used for household lighting. Such bulbs are rated according to their power, 60, 100 W, and so on. For industrial-scale wind power, larger units are necessary—kilowatt (1 kW = one thousand W), megawatt (1 MW = one million W), and gigawatt (1 GW = one billion W). Large, modern wind turbines typically have power capacities in the range 1.5 to 3.0 MW (approx. 2,000–4,000 hp) or greater.

Energy, on the other hand, is power applied over a period of time. The most common unit is the kilowatt-hour (kWh), in other words, one kilowatt used continuously over the period of one hour. For example, burning a 100-W light bulb for 10 hours represents 1 kWh. This is the way that electricity is metered and the means by which people are typically charged for their electricity consumption.

Wind power (P) is determined by three factors, namely air density (ρ), the area swept by the turbine rotor (A), and the cube of instantaneous wind velocity (V^3) according to the following formula (Musgrove 2010):

$$P = \tfrac{1}{2} \rho A V^3$$

The cube of V is the most critical factor, as first demonstrated by la Cour in the late 1800s (see sect. 1.3). For instance, doubling velocity increases the power by eight times ($2^3 = 8$). Also as noted, wind velocity depends, in turn, on surface roughness (α) and height (H) above the ground. Over much of the Great Plains in North America, for example, shortgrass prairie predominates, and the roughness exponent is approximately 1/7 (0.14). In this situation, raising the tower height from 10 to 60 m increases mean annual wind speed by more than a quarter, which more than doubles the power (Musgrove 2010). Height above the ground is, in fact, a critical factor for harnessing wind energy (Gipe 2009).

Increasing the diameter of the turbine rotor also leads to greater power; doubling rotor diameter increases the swept area (A) fourfold. Hence, the trend through time for taller towers and larger rotors. Modern, large wind turbines typically have tower heights of 80–120 m and rotor diameters of 90 to 120 m. Turbine manufacturers continue to push the height and rotor-diameter limits to achieve still greater power capacity and, thus, more efficient machines. The Vestas V236-15 MW offshore turbine, for example, has blades 115.5 m long and a rotor diameter of 236 m (Vestas V236 2025).

In the United States, the limiting height factor is Federal Aviation Adminstration (FAA) restrictions for structures that extend into the national airspace. In general, any structure standing more than 200 feet (61 m) above the ground requires FAA approval. This is usually not a problem up to 500 feet (152 m) in height, which is normally the lower flying limit for manned aircraft. For many years this was the de facto height limit for turbines in the United States. However, 500 feet is no longer an absolute height limit. According to current FAA guidelines for wind turbines:

> *... there is no restriction on heights for filing notice. ... Please note that at 500 feet or higher, your proposed structure will be in altitudes available to general aviation aircraft. A public notice may be issued to gather aeronautical information ... Only valid aeronautical comments will be factored into the FAA's decision. Example: Comments concerning land values or aesthetics would not be considered valid aeronautical comments.* (FAA 2023)

Taller turbines, in the range 160 to 200 m height, have appeared in many U.S. wind farms since 2018 (Fig. 3.8; Hoen et al. 2018–2025). However, individual states and counties may have their own height restrictions. In Massachusetts, for instance, small turbines are limited to 250 feet height, and the limit for large turbines is 450 feet, although exceptions for taller turbines are allowed (Massachusetts Energy 2011).

Lateral spacing of turbines on the ground is also an important consideration, as the downwind turbulence from one turbine could affect neighboring turbines. Spacing of turbine rows is typically in the range of 5 to 10 times tower height or seven times rotor diameter (Twidell 2009). For example, turbine towers 90 m tall

Figure 3.8 Vestas V120-2.2 MW turbines have a rotor diameter of 120 m and total height of 182 m (nearly 600 feet) with a blade in the upright position. North Fork Ridge wind farm, southwestern Missouri. Hoen et al. (2018–2025)/Kite photograph (2022).

would be spaced approximately 600 m apart. However, this spacing arrangement may be varied depending on local topography, land use, prevailing wind, and other factors. Given the land-survey system in the central and western United States, which is based on square-mile sections, wind turbines are often spaced in lines half a mile (~800 m) apart along existing field boundaries (see Figs. 1.19 and 2.3). This arrangement minimizes loss of agricultural land use.

3.3 Turbine Efficiency and Capacity

Rotor swept area (A) and the average annual wind speed at hub height (V) are the two most important factors for determining wind power available for potential electricity generation, as noted above. However, conventional turbines cannot

Figure 3.9 NedWind 40/500 two-bladed turbines. Rotor diameter is 40 m, hub height is 39.5 m, and capacity is 500 kW. NedWind began in 1990 in the Netherlands and was taken over by NEG Micon in 1998, which was acquired in turn by Vestas in 2004 (Bauer and Matysik 2025a). Garnet Wind Project was constructed in 2009 in the San Gorgonio Pass near Palm Springs, California. Hoen et al. (2018–2025)/USGS/Public Domain.

capture all the power that is available in the wind. The Betz limit is the theoretical maximum that ideal turbine rotors could achieve with no aerodynamic or mechanical losses and which allows for the wind to pass through the turbine. Named after the German aeronautical engineer, Albert Betz, who developed the concept in the 1920s, the power coefficient (C_P) for perfect rotor operation is 59% (Musgrove 2010).

Well-designed blades of modern turbines may capture ~40% or slightly more, but this figure applies only to the rotor. Additional losses take place in the mechanical components and generator of the turbine, which lead to an overall conversion efficiency of about one-third for most large turbines (Gipe 2009). Smaller turbines generally have lower efficiencies of only 20% or less.

Modern wind turbines typically have three rotor blades or rarely two blades (Fig. 3.9). The frontal surface area of the blades in comparison to the swept area (A) is known as the solidity of the rotor. In modern wind turbines this is a relatively small value, usually less than 5% (Musgrove 2010). Traditional windmills have solidity in the 15–30% range (see Figs. 1.2 and 1.3), and American windmills are usually ~70% (see Fig. 1.8). Manufacturing long, thin, fiberglass blades for modern turbines is, however, quite expensive, so three blades with low rotor solidity have become the norm. Two blades are more common for smaller turbines (Fig. 3.10).

Figure 3.10 Gaia-Wind small turbine near Heltborg, northwestern Denmark. Wind blowing from left to right. Two-bladed, 11 kW capacity turbine with 13-m-diameter rotor mounted downwind from the tower (Huskey et al. 2010). The two blades are made of a single piece of fixed fiberglass with centrifugally activated flap brakes near the blade tips. Gaia-Wind was founded in Denmark in 1993, later moved to Scotland, and is now known as Ryse Energy.

A wind-turbine blade has much in common with an airplane wing. In both cases, the airfoils have low drag compared with the lift they produce. For turbine blades, a key factor is the tip-speed ratio (λ), which is the relationship between blade-tip speed and upstream wind velocity (V). The blade-tip speed depends, in turn, on blade angular velocity (ω) and length (radius) of the blade (R), such that:

$$\lambda = (\omega \times R) / V$$

For large, three-bladed turbines, the optimum λ value is considered to be ~7 with reported values in the range 6–8, which result in C_P of 40–45% (Musgrove 2010; Ragheb 2014). For a given wind speed, turbines with shorter blades must rotate faster, whereas large turbines rotate slower to achieve the same λ value (Fig. 3.11). This has important implications for environmental issues, noise, and aesthetic appearance (see Chap. 5).

Figure 3.11 Comparison of rotor diameter, tip-speed ratio (λ), and wind velocity. Vestas V47 vintage 2001 (left) and Siemens 2.3 MW vintage 2012 (right). Blade length (*R*) for the Vestas turbine = 23.5 m, and *R* = 54 m for the Siemens turbine. To achieve the same λ value for a given wind velocity, the Vestas turbine would have to rotate more than two times faster than the larger Siemens model. Wind farms in southwestern Kansas.

The nacelle is the housing atop the tower in which blade rotation is converted into electricity. The blades are affixed to a hub that connects to a gearbox in which the rate of rotation is stepped up to the generator (Fig. 3.12). Most large turbines employ induction generators that may operate at partial variable speed and which may be connected directly to the electrical grid system (Gipe 2009), although many variations exist on this basic setup that are beyond the scope of this book. For more advanced discussions of wind-turbine design and engineering, readers are advised to consult Manwell et al. (2020) or Burton et al. (2021).

Wind turbines are typically given a power rating for a specific wind speed at which the turbine reaches its maximum output. For instance, the Vestas V100 1.8 MW turbine achieves its rated 1.8 MW capacity at 12 m/s wind speed, but produces only 1.0 MW at 8 m/s (Fig. 3.13). Actual power production in routine operation is known as the capacity factor, which is the total energy output measured over an extended time interval, usually a year, divided by the potential rated output for the same period (Musgrove 2010).

Figure 3.12 Final assembly of Turbowinds T600-48 mid-sized turbine showing the hub (A), main bearing (B), main shaft (C), gearbox (D), brake (E), induction generator (F), nacelle frame (G), and cover (H). This model is rated 600 kW, rotor diameter is 48 m, tower height is 50–60 m, and output voltage is 690 V. Turbowinds was founded in Belgium in 1985, but discontinued in 2018; the T600-48 was the largest model (Bauer and Matysik 2025b). TURBOWINDS NV/Wikimedia Commons/CC BY-SA 4.0

Figure 3.13 Vestas V100-1.8 MW turbines were erected in 2012 in the Busch Ranch wind farm of southeastern Colorado. Rotor diameter is 100 m, and total height is 130 m. Hoen et al. (2018–2025).

Figure 3.14 Overview of the Lamar Light and Power wind turbines situated south the Arkansas River valley in southeastern Colorado. This small array consists of four GE Wind 1.5 MW turbines and was constructed in 2004. Rotor diameter is 70.5 m, hub height is 80 m, and total height is 115.2 m. Hoen et al. (2018–2025). Panorama assembled from two, overlapping, kite aerial photographs.

The capacity factor is usually expressed as a fraction or percentage, which may be calculated for a single turbine or, more commonly, for multiple turbines in a wind farm. As an example, the four GE Wind turbines operated by Lamar Light and Power on the High Plains in southeastern Colorado have a long-term capacity factor of 33%, and their reliability is greater than 95% (Fig. 3.14; Lamar 2025). In more favorable topographic situations, the capacity factor may exceed 40%. Newer, larger offshore turbines have even higher capacity factors; for instance, the Vestas V236 turbine boasts a capacity factor that exceeds 60% (Vestas V236 2025).

Evaluation of such large turbines takes place in Denmark at Høvsøre on the western coast of Jutland at a test facility operated by Risø DTU, the National Laboratory for Sustainable Energy at the Technical University of Denmark. Wind from the North Sea blows without interruption across the flat terrain, where it is measured continuously at multiple heights by meteorological towers in proximity to turbines, and is compared with actual turbine performance (see Fig. 4.4). Similar testing of small wind turbines takes place at the Nordic Folkecenter for Renewable Energy (Fig. 3.15). Other institutions for testing and research include Lindø Offshore Renewable Center (LORC) in Denmark and the U.S. National Renewable Energy Laboratory (NREL 2025).

Figure 3.15 Small turbines are tested at the Nordic Folkecenter for Renewable Energy near Hurup, northwestern Jutland, Denmark. These turbines are designed for individual homes, small businesses, or farms with capacities ≤75 kW.

Turbine Decommissioning

Wind turbines are generally designed for a lifespan of 20–25 years. Already many older, small turbines in Denmark and California have been taken down or replaced by larger, newer turbines. Older wind farms throughout the U.S. Great Plains are beginning to show their age. These turbines eventually must be dismantled and removed. Decommissioning of old wind turbines already has begun, for example, in North Dakota (Anderson 2023).

Most turbine components—steel, copper, electronics—could be recycled. However, the long fiberglass blades are more difficult to handle (Coleman 2023). A new company in Iowa has developed a proprietary process for recycling turbine blades as well as new turbine-blade manufacturing scrap material into reinforcing fibers for concrete, asphalt, and other construction materials (REGEN 2024). Reuse is another approach for dismantled components of wind turbines. Many older Danish turbines dating from the 1980s in California have been rebuilt and installed across the country for farms and small businesses.

Vattenfall, a Swedish energy company, has produced a prototype model of a tiny house made from a turbine nacelle shell. The nacelle was derived from a Vestas V80-2.0 MW turbine that served in the Austrian Gols wind farm for 20 years. Tens of thousands of nacelles this size or larger will become available in the future as they are retired. This is an innovative way to reuse a sizable turbine component, rather than energy-intensive melting and reforging the steel alloy. According to Thomas Hjort, this *saves raw materials, energy consumption, and in this way we ensure that these materials are useful for many years after their first working life* (Vattenfall 2024).

Tehachapi wind-energy complex as seen in 2012. Small turbines on truss towers date from the early 1980s, and some towers are derelict. Many of these were taken down later and replaced by fewer, larger turbines on tubular towers. Near Oak Creek Pass, Tehachapi Mountains, California.

Turbine Decommissioning (Continued)

Taking down a 20-year-old Nordex wind turbine in North Dakoka in 2022. The base of the tower was cut with torches and then pulled down. The structure was dismantled for disposal or recycling. Note the blade-tip brake flaps (compare with Fig. 1.17). Courtesy of J. Schramm of The Minot Daily News.

TalkInc turbine at a farm near Waverly, Kansas. TalkInc turbines are remanufactured in Minnesota from previous equipment. This one is derived from a Windmatic model, which was built originally in Denmark during the 1980s.

3.4 Electricity Grid and Interconnection

For a single wind farm or local area, the wind varies daily and seasonally, which explains why capacity factors are usually in the 30–60% range. Simply put, the wind is not constant at the ideal speed for turbines to produce continuously their maximum rated output. Periods of light wind or calm happen from time to time, when turbines shut down (Fig. 3.16). To opponents of wind energy, this has been a convenient argument against further harnessing of wind resources, which some people regard as undependable. However, this argument ignores the regional nature of wind energy and the grid systems used to distribute electricity over long distances to consumers.

In North America, electricity is distributed within linked grids of transmission lines, known as interconnections. Thus, electricity generated in multiple locations by several means may be transferred via the grid system to consumers in many other places. In this way, the dispersed nature of wind energy overcomes the vagaries of local weather and wind conditions (Aber et al. 2015).

Figure 3.16 Turbines not operating in low-wind conditions. Turbine rotors face in different directions, and the blades are feathered vertically parallel to the towers. Sany 2.0 MW turbines in the Huerfano River Project on the High Plains near Walsenburg, Colorado Hoen et al. (2018–2025).

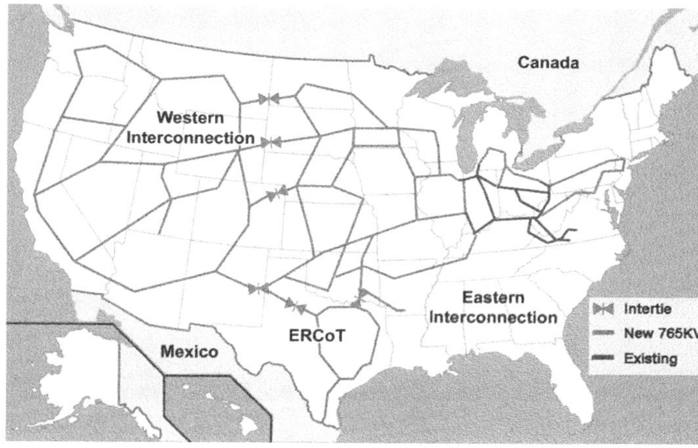

Figure 3.17 Super electrical grid using 765 kV AC transmission lines and interties. Proposed for the United States in 2008 by the American Wind Energy Association. Wikimedia Commons https://commons.wikimedia.org/

As weather systems move across the Great Plains of North America, for instance, wind speed and direction shift so that some locales may have weak or calm conditions, while others places have strong wind. Throughout the region some wind farms are generating at peak capacity, while others are operating at partial capacity, and a few are not generating on any particular day. Overall regional wind-power production continues to feed into the grid system to be used where needed. Three interconnections serve the contiguous (48) United States and parts of Canada and Mexico (Fig. 3.17).

- Western Interconnection—Pacific and Rocky Mountain states as well as western Canada and part of Mexico, represented by the Western Electric Coordinating Council.
- Eastern Interconnection—Atlantic coast to the Great Plains in the United States and south-central Canada, including several regional councils and reliability organizations.
- Electric Reliability Council of Texas (ERCoT)—separate interconnection serving most of the state of Texas.

Within each interconnection, alternating current (AC) power is synchronized, so that electricity may be transferred from diverse energy sources to numerous consumers. In this way, electric power may be distributed as needed from redundant sources via multiple long-distance pathways with better overall efficiency. The key to effective movement of electricity is high-voltage transmission lines (Fig. 3.18).

The proposed supergrid infrastructure plan (Fig. 3.17) was killed by the first Trump administration in order to help the coal industry. The Trump effort to

Figure 3.18 Three sets of high-voltage electricity transmission lines cross paths here in the Flint Hills of east-central Kansas.

support coal was largely unsuccessful; however, the lack of an integrated nationwide supergrid has continuing consequences. Wind-generated electricity from western Kansas, for example, cannot be shared easily with the large urban/industrial area along the Front Range of the Rocky Mountains in Colorado. Texas suffered an energy calamity in February 2021 because of its isolated situation.

Texas Energy Fiasco

The energy infrastructure of Texas suffered its worst failure in February 2021 as a result of a series of severe winter storms. More than 4.5 million homes and businesses lost power; there were shortages of water, food, and heat, and about 250 people died as direct or indirect results. Texas Governor Abbott and others initially blamed frozen wind turbines and solar panels, which was utterly untrue.

Hindsight has demonstrated the primary cause for this calamity was due largely to the failure of natural-gas-powered generators (Homeland Security 2021). This led to partial shutdown of the electric-grid system by the Electric Reliability Council of Texas (ERCoT), which is independent of other electric-grid interconnections in the United States. The disconnection of ERCoT made it difficult, in fact nearly impossible, to import electricity from outside the state.

Texas had been warned a decade before that its electric-grid system was vulnerable to failure during cold weather. But this warning went unheeded. In fact, cold weather had caused previous system-wide rolling blackouts, most

Texas Energy Fiasco (Continued)

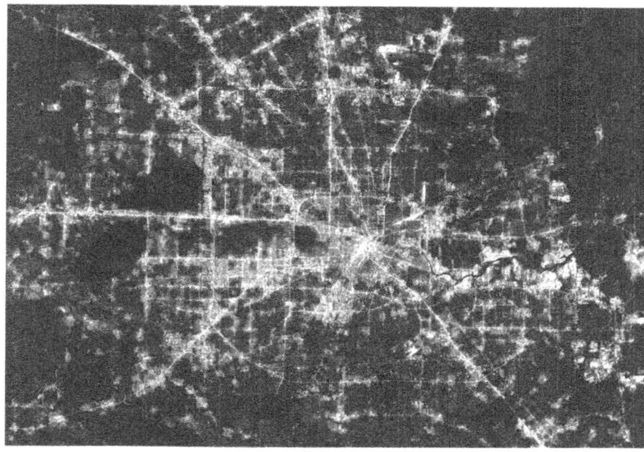

Night-time view of Houston, Texas from February 7, 2021. Note numerous dark areas within the city grid. NASA composite image derived from Landsat 8 and Suomi NPP satellites. Stevens and Hansen (2021)/ NASA/ Public Domain.

Vestas V100-2.0 turbine standing in a wheat field in the Lindahl wind farm of northwestern North Dakota. Hoen et al. (2018–2025). Winter temperature in this vicinity may drop below −30 °C.

(Continued)

Texas Energy Fiasco (Continued)

notably in 1989 (Homeland Security 2021), long before wind energy came to Texas. So, the potential for cold-weather impact on ERCoT was well known.

Wind turbines and natural gas certainly may be prepared for operation in extreme cold conditions, such as the northern Great Plains. Wind turbines are particularly abundant in Minnesota and North Dakota, states known for bitter cold winters. Nickel–stainless-steel alloys are used for key components along with other cold-climate options for low-temperature turbine operation (see Chap. 9). Typical Vestas turbine models, for instance, are rated for operation down to −30 °C and for withstanding ambient temperature as low as −40 °C (Vestas Cold 2019).

Much finger-pointing and disinformation was circulated at the time in Texas. For instance, a picture of a helicopter de-icing turbine blades, supposedly from Texas, turned out to be an experiment in Sweden. But the simple truth is that ERCoT neglected to prepare its electrical infrastructure for severe cold weather conditions, in spite of warnings and previous failures. Such preparations are routine elsewhere and really have nothing to do with the potential for wind energy. The go-it-alone attitude in Texas also contributed to this disastrous power calamity.

President Biden's Bipartisan Infrastructure Law was passed and signed in 2021 and has led to a dramatic change in how the United States approaches energy challenges of the twenty-first century. In 2022, for instance, the Department of Energy (DOE) released $28 million for wind-energy projects (DOE Wind 2022):

- High-voltage direct current (DC) for offshore wind transmission.
- Advancing deployment of distributed wind.
- Offshore wind-energy social science research.
- Bat deterrent technology development.

As these topics demonstrate, federal funding deals with engineering issues as well as social and environmental aspects for deploying wind energy. The Inflation Reduction Act of 2022 further supports development of clean energy through tax credits and consumer rebates. The bill includes 950 million solar panels, 120,000 wind turbines, and 2,300 grid-scale battery plants (White House 2022).

The DOE Office of Electricity is promoting grid modernization and the smart grid under authority of the Energy Independence and Security Act of 2007. Americans for a Clean Energy Grid have identified 22 transmission projects designed to access wind and solar energy and to interconnect about 60 GW of

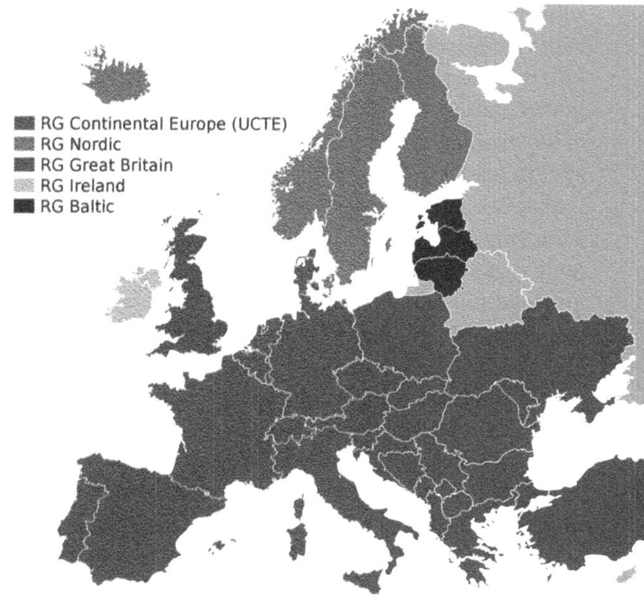

Figure 3.19 Electricity transmission regional groups (RG) for western and central Europe. Public domain map obtained and adapted from Wikimedia Commons https://commons.wikimedia.org/

renewable capacity (Goggin et al. 2021). These are feasible projects that could be built within this decade and would join Eastern, Western, and ERCoT interconnections. At the time of writing this book, however, the election of President Trump to a second term has led to changes in U.S. energy policy.

Similar efforts are underway in Europe. In fact, the synchronous grid of continental Europe (UCTE) is interconnected in a single phase-locked 50 Hz frequency system (Fig. 3.19). It provides electricity to more than 400 million people in 24 countries including most of the European Union (EU). In 2018, the EU set an electricity interconnection target of 15% by 2030 (EU Interconnection 2024). In other words, each country should have interconnections sufficient to share up to 15% of electricity generated on its territory with neighboring countries. This target was established specifically to deal with increasing wind and solar capacities and their variability. As of 2021, 16 countries were on track to reach the 15% target by 2030 or had already attained that level. But more interconnections are needed in some regions.

3.5 Integration of Wind Energy

Smart-grid technology would be utilized to monitor production, transmission, and consumption of electricity as grid systems are upgraded and new interconnections are built both in Europe and North America. Wind forecasting substantially enhances the operation of such grid systems (DOE Wind 2008). Wind forecasts provide the basis for estimating wind-generated electricity feeding into the grid system. Other generating sources, such as coal, gas, nuclear, or hydro power plants as well as other types of renewable energy, could be adjusted upward or downward to compensate for fluctuations in wind energy in order to meet system load requirements.

Seamless integration of all types and sources of electricity is a critical step toward achieving substantial wind-energy contribution to the overall electricity needs of a large region or country. The electric utility operator needs to analyze the regulation and load-following requirements for wind relative to other generation sources. For example, both wind and solar energy generate electricity primarily during daylight hours, when most electricity is consumed (Fig. 3.20). While this concept may seem simple, it is exceedingly complex to accomplish in practice.

The power available in the wind is an energy resource, but only a portion of its capacity is available at any given moment. Furthermore, wind energy cannot be scheduled to meet human needs on a daily or seasonal basis. Conventional power plants, on the other hand, represent capacity resources that may be scheduled and adjusted as needed (DOE Wind 2008). A gas-fired generating station, for example, may adjust its output on demand to meet peak load requirements. Existing electrical supply and distribution systems have been designed to ensure reliable capacity and to meet peak load conditions, but each of the generating sources serves different functions.

Wind energy has been added primarily to provide electricity from renewable and non-polluting resources, but it cannot replace capacity resources to meet peak load requirements. In order to overcome the variability of wind and solar power, battery-storage facilities are under development. Lithium-ion batteries store excess renewable wind and solar energy, then release the electricity when needed

Figure 3.20 Sunpower solar-energy facility in the San Luis Valley, near Alamosa, Colorado. Panels are tilted at a 20° angle to the southern horizon and rotate during the day to track the sun.

Figure 3.21 Double set of overhead AC transmission lines near Agerskov in southwestern Denmark. This is a main electrical-grid connection between Jutland and Germany in the UCTE.

to transmission lines. Typical plants have capacities of 200 MW and up to 800 MWh, which is enough electricity to power 200,000 homes for four hours. Some 2,300 grid-scale battery plants are planned in the United States, as noted above.

Denmark remains at the forefront of renewable energy development. In general, the country generates more electricity than it needs, primarily from wind; however, at any given moment Denmark may export electricity to some countries and import electricity from other countries in order to balance its system load between wind and solar energy and central power stations (Fig. 3.21). Construction of two energy islands, an artificial one in the North Sea and another on Bornholm in the Baltic Sea, promises to further integrate Denmark by the 2030s with surrounding countries—Germany, Netherlands, Norway, Poland, Sweden, and United Kingdom (see sect. 6.2.4).

3.6 Summary

Wind is the atmosphere in motion, which is a consequence of air pressure differences. Air pressure, in turn, is largely a function of temperature differences; surface wind represents the transfer of air masses from high-pressure zones of cold sinking air toward low-pressure zones of warm rising air. Global wind patterns, such as the trade winds and westerlies, tend to be quite stable over large oceans, but are more variable over land, especially the interiors of large continents. The positions of major geographic features influence further complications in local and regional wind patterns, which may have distinct diurnal rhythms and seasonal shifts. Diurnal wind usually reaches peak velocity during the afternoon hours for most locations. Air pressure declines with altitude, so turbine makers have to take the thin air at high altitude into account for turbine design.

Near-surface turbulent flow dissipates wind power, so turbine height above the ground is a critical factor for producing wind energy. Increasing the size of the rotor is another means to boost turbine output; hence, the trend through time for taller towers and larger rotors. Wind turbines are capable of intercepting and harnessing only a portion of the available wind power; this conversion efficiency is typically about one-third for most large turbines.

Furthermore, wind turbines do not operate continuously at peak output, because of variations in wind speed. Actual power production in routine operation is known as the capacity factor, and is usually in the range 30–40% and may exceed 60% for large, offshore wind farms. The terms energy and power are often used interchangeably. But strictly speaking, power is the rate at which energy is utilized, and energy is power applied over a period of time. Common electrical units are the kilowatt (kW) for power and kilowatt-hour (kWh) for energy.

The dispersed nature of wind energy, spread over large geographic regions, overcomes the vagaries of local weather. Energy generated by numerous wind farms feeds into regional grid systems that transfer electricity via interconnections to distant consumers. Wind forecasting enhances the operation of such grid systems substantially. Existing electrical supply and distribution systems have multiple, redundant sources, each of which serves different functions. Wind energy has been added primarily to provide electricity from renewable and non-polluting resources, and battery-storage facilities will help to even out the variability of wind and solar electricity production.

References

Aber, J.S., Aber, S.W. and Pavri, F. 2015. Windscapes: A global perspective on wind power. Multi-Science Publ., UK, 245 p.

Aber, J.S., Marzolff, I., Ries, J.B. and Aber, S.E.W. 2019. Small-format aerial photography and UAS imagery: Principles, techniques and geoscience applications, 2nd ed. Elsevier, Amsterdam, Netherlands, 382 p.

ALMA 2025. Atacama large millimeter/submillimeter array: Climate. https://almascience.eso.org/about-alma/weather/overview

Anderson, F.J. 2023. Geologists see continued growth in civil and energy construction activity in North Dakota. *North Dakota Mineral Resources, Geo News* 50(1), p. 17–21. https://www.dmr.nd.gov/ndgs/documents/newsletter/2023Winter/Geologists_See_Continued_Growth_in_Civil_and_Energy_Construction_Activity_in_North_Dakota.pdf

Bauer, L. and Matysik, S. 2025a. NedWind Rhenen bV. Wind-turbine-models. https://en.wind-turbine-models.com/manufacturers/74-nedwind

Bauer, L. and Matysik, S. 2025b. Turbowinds SA/NV. Wind-turbine-models.com. https://en.wind-turbine-models.com/manufacturers/120-turbowinds

Bernardos, A. and Fulda, J. 2020. Why wind energy is quite alright: Examining 6 common arguments against wind energy. Medium. https://medium.com/clir-tech-blog/why-wind-energy-is-quite-alright-d671ff40194a

Bloomfield, H. 2021. What Europe's exceptionally low winds mean for the future energy grid. The Conversation. https://theconversation.com/what-europes-exceptionally-low-winds-mean-for-the-future-energy-grid-170135

Burton, T., Jenkins, N., Bossanyi, E., Sharpe, D. and Graham, M. 2021. Wind energy handbook, 3rd ed. Wiley, 960 p.

Chen, L. 2020. Impacts of climate change on wind resources over North America based on NA-CORDEX. *Renewable Energy* 153, p. 1428–1438. https://doi.org/10.1016/j.renene.2020.02.090

Cohen, J.E. and Small, C. 1998. Hypsographic demography: The distribution of human population by altitude. *Proceedings of the National Academy of Sciences, Applied Physical Sciences, Social Sciences* 95, p. 14009–14014. http://www.pnas.org/content/95/24/14009.full.pdf

Coleman, K. 2023. Turbine recycling advancements. *Colorado County Life* 54(4), p. 14. https://issuu.com/coloradocountrylife/docs/april_2023_kc/14

CSU. 2013. Colorado anemometer long program: Florissant – 11/17/2011 to 2/5/2013. Colorado State University. https://projects-web.engr.colostate.edu/ALP/ALP_100_Florissant.html

Dizikes, P. 2010. Energy answer: Blowing in the wind? MIT News. http://web.mit.edu/newsoffice/2010/wind-economics-0525.html

DOE Wind 2008. 20% wind energy by 2030: Increasing wind energy's contribution to U.S. electricity supply. U.S. Department of Energy, DOE/GO-102008-2567. http://www.nrel.gov/docs/fy08osti/41869.pdf

DOE Wind 2022. WETO releases $28 million funding opportunity to address key deployment challenges for offshore, land-based, and distributed wind. U.S. Department of Energy, Wind Energy Technologies Office. https://www.energy.gov/eere/wind/articles/weto-releases-28-million-funding-opportunity-address-key-deployment-challenges

DWIA. 2003. Wind speed variability: Diurnal (night and day) variations of the wind. Danish Wind Industry Association.

EU Interconnection 2024. Electricity interconnection targets. European Commission: Energy. https://energy.ec.europa.eu/topics/infrastructure/electricity-interconnection-targets_en

FAA 2023. Wind turbine FAQs (latest revision 10/19/2023). Federal Aviation Administration. https://oeaaa.faa.gov/oeaaa/external/searchAction.jsp?action=showWindTurbineFAQs

Gipe, P. 2009. Wind energy basics: A guide to home- and community-scale wind energy systems, 2nd ed. Chelsea Green Pub, White River Junction, Vermont, 166 p.

Goggin, M., Gramlich, R. and Skelly, M. 2021. Transmission projects ready to go: Plugging into America's untapped renewable resources. Americans for a Clean Energy Grid. https://cleanenergygrid.org/wp-content/uploads/2021/09/Transmission-Projects-Ready-to-Go.pdf

Graetz, M.J. 2011. The end of energy: The unmaking of America's environment, security, and independence. MIT Press, Cambridge, Massachusetts, 369 p.

Hoen, B.D., Diffendorfer, J.E., Rand, J.T., Kramer, L.A., Garrity, C.P. and Hunt, H.E. 2018. United States Wind Turbine Database V8.0 (February 25, 2025). U.S. Geological Survey, American Clean Power Association, and Lawrence Berkeley National Laboratory data release. https://energy.usgs.gov/uswtdb/ or https://doi.org/10.5066/F7TX3DN0

Homeland Security 2021 Texas power crisis. Center for Homeland Defense and Security. https://www.hsdl.org/c/tl/2021-texas-power-crisis/ [site removed by Trump administration]

Huskey, A., Bowen, A. and Jager, D. 2010. Wind turbine safety and function test report for the Gaia-Wind 11-kW wind turbine. U.S. National Renewable Energy Laboratory, Technical Report, NREL/TP-500-46724. https://www.nrel.gov/docs/fy10osti/46724.pdf

Lamar 2025. Generation portfolio. Lamar Light and Power. https://www.lamarlightandpower.com/generation-portfolio/

Li, M. and Li, X. 2005. Investigation of wind characteristics and assessment of wind energy potential for Waterloo region, Canada. *Energy Conversion and Management* 46(18–19), p. 3014–3033. https://doi.org/10.1016/j.enconman.2005.02.011

Lledó, L., Bellprat, O., Doblas-Reyes, F.J. and Soret, A. 2018. Investigating the effects of Pacific Sea surface temperatures on the wind drought of 2015 over the United States. *Journal of Geophysical Research – Atmospheres* 123, p. 4837–4849. https://doi.org/10.1029/2017JD028019

Manwell, J.F., McGowan, J.G. and Rogers, A.L. 2020. Wind energy explained: Theory, design and application, 2nd ed. Wiley, 690 p.

Massachusetts Energy 2011. Model amendment to a zoning ordinance or by-law: Small wind energy systems. Massachusetts Division of Energy Resources, Massachusetts Executive Office of Energy and Environmental Affairs https://www.mass.gov/doc/model-allow-wind-smallpdf/download

Massen, F., Kies, A. and Harpes, N. 2007. Seasonal and diurnal CO_2 patterns at Diekirch, LU, 2003–2005. MeteoLCD Paper 45. https://meteo.lcd.lu/papers/

Musgrove, P. 2010. Wind power. Cambridge University Press, 323 p.

NGMD 2025. The national geologic map database. U.S. Department of the Interior. https://ngmdb.usgs.gov/ngmdb/ngmdb_home.html

NREL 2025. Wind research. U.S. National Renewable Energy Laboratory. https://www.nrel.gov/wind/research

Olsson, L.E. 1994. Energy-meteorology: A new discipline. *Renewable Energy* 5, p. 1243–1246. https://doi.org/10.1016/0960-1481(94)90157-0

Ragheb, M. 2014. Optimal rotor tip speed ratio. https://user-web-p-u02.wpi.edu/~cfurlong/me3320/DProject/Ragheb_OptTipSpeedRatio2014.pdf

Reddy, G.G.K., Reddy, S.V., Ramkumar, T.K. and Sarojamma, B. 2014. Wind power density analysis for micro-scale wind turbines. *The International Journal of Engineering and Science* 3, p. 53–60.

REGEN 2024. Press release: REGEN fiber unveils innovative wind turbine recycling solution. REGEN Fiber. https://regenfiber.com/

Stevens, J. and Hansen, K. 2021. Extreme winter weather causes U.S. blackouts. NASA Earth Observatory. https://earthobservatory.nasa.gov/images/147941/extreme-winter-weather-causes-us-blackouts?src=ve

Twidell, J. 2009. Fundamentals of wind turbines. In: J. Twidell and G. Gaudiosi (eds.), Offshore wind power. Mulit-Science Publ. Co., Brentwood, Essex, United Kingdom, p. 15–26.

Vattenfall 2024. Wind turbine turned into compact living. https://group.vattenfall.com/press-and-media/newsroom/2024/wind-turbine-turned-into-compact-living

Vestas Cold 2019. Vestas cold climate solutions. Vestas. https://pdf.archiexpo.com/pdf/vestas/vestas-cold-climate-solutions/88087-377614.html

Vestas V236 2025. V236-15.0 MW™. https://www.vestas.com/en/energy-solutions/offshore-wind-turbines/V236-15MW

White House 2022. By the numbers: The Inflation Reduction Act. White House Briefing Room. https://www.whitehouse.gov/briefing-room/statements-releases/2022/08/15/by-the-numbers-the-inflation-reduction-act/ [site removed by Trump administration].

Yan, P., Zuo, D., Yang, P. and Li, S. 2021. Typical modes of the wind speed diurnal variation in Beijing based on the clustering method. *Frontiers in Physics* 9, 675922. https://www.frontiersin.org/articles/10.3389/fphy.2021.675922/full

4

Public Policy for Wind Energy

4.1 Introduction

There is widespread evidence of humans harnessing wind energy from early civilizations through the onset of the modern industrial era. However, the ascendancy of coal, oil, and other consistent energy sources during the Industrial Revolution precipitated a decline in wind-energy usage. These alternatives offered greater reliability, ease of transport, and economies of scale that wind energy could not effectively match. In the United States, the expansion of power grids across vast swaths of rural areas during the twentieth century further cemented the decline of small-scale, stand-alone, wind-energy development and its use across even rural landscapes (USEIA 2023).

In recent decades, wind energy has experienced a significant resurgence globally—a revival few could have anticipated in the mid-twentieth century. This resurgence is driven by several generally known factors. First, the geopolitical vulnerabilities and price volatility associated with fossil fuels have prompted nations to seek more stable and indigenous energy sources. Wind energy, being inexhaustible, enhances energy security by reducing dependence on imported fuels. Second, the imperative to address climate change has driven the widespread adoption of renewable energy sources. Wind power, as a clean energy source, plays a vital role in reducing greenhouse gas emissions.

Third, innovations in turbine technology and economies of scale have significantly reduced the cost of wind energy, making it more competitive with traditional energy sources. Fourth and perhaps most importantly, political will, the right mix of favorable renewable energy policies, production incentives, trading networks, affordable financing, and tax credits implemented across different countries have worked to extend the markets, capacity, and infrastructure

Wind Energy Landscape: Principles, Techniques, History, First Edition. James S. Aber, Susan E. W. Aber, and Firooza Pavri.

Published 2026 by John Wiley & Sons, Inc.

for wind energy. Collectively, these developments position wind energy as a promising contributor to diversifying national energy portfolios in the twenty-first century.

Governments outline broad strategies and frameworks for the development and management of natural resources. Setting public policy in the arena of natural and energy resources is an inherently political process shaped by the broader public discourse of the day. Public policy in the wind-energy arena has a relatively recent history when compared to other natural resources like forests or even wetlands. While historical and cultural contexts influence a country's approach to wind energy, the field is in its developmental stages. Inconsistencies or shifts in national policies, regulatory frameworks, and market sizes contribute to variability in the stability and trajectory of wind-energy development (Lewis and Wiser 2007). Even so, the dramatic increase in global cumulative installed wind power capacity and investments in offshore wind projects underscore the dynamic nature of the wind-energy sector and the critical role of policy in shaping its future.

4.2 Wind-Energy Policy Measures

Specific policy measures or policy instruments, as they are also known, have driven significant growth of wind-energy production worldwide. Large-scale developments in alternative energy sectors, particularly wind and solar power, are not yet fully competitive with conventional fossil fuels in terms of cost and scalability. Without the support of targeted policy interventions, it would be challenging to sustain the continued expansion of installed wind-energy capacity observed since the late 1990s.

Data from 2023 for the top wind-energy producing nations suggest as much, with increases accounted for by favorable policies and political will (Fig. 4.1). Global, cumulative, installed wind power capacity crossed 1,000 GW in 2023 (IRENA 2024). In wind-energy leader China, for example, total installed capacity increased from approximately 0.36 GW in 2,000 to 29.63 GW in 2010 and 441.89 GW by 2023. Meanwhile, in the United States over a 10-year period installed capacity increased from 64.43 GW in 2014 to 140 GW in 2023. Increases seen in China, the United States, and elsewhere have come hand in hand with advancements in turbine technology, economies of scale, and critical policy support, including subsidies and incentives (Fig. 4.2; Saidur et al. 2010).

Energy strategies promoted by governments often include a mix of legislation, direct and indirect economic and financial incentives, taxation policies, financing mechanisms, and research and development (R&D) support. Countries rarely adopt just one policy instrument for any given targeted industrial sector. Rather they select from a varied mix of measures based on their alignment with

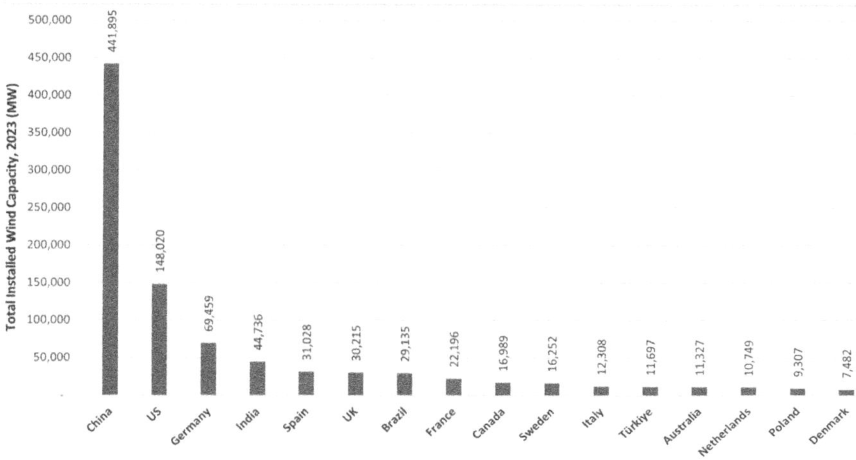

Figure 4.1 Cumulative total installed wind-energy capacity (MW) for selected countries, 2023. IRENA (2024).

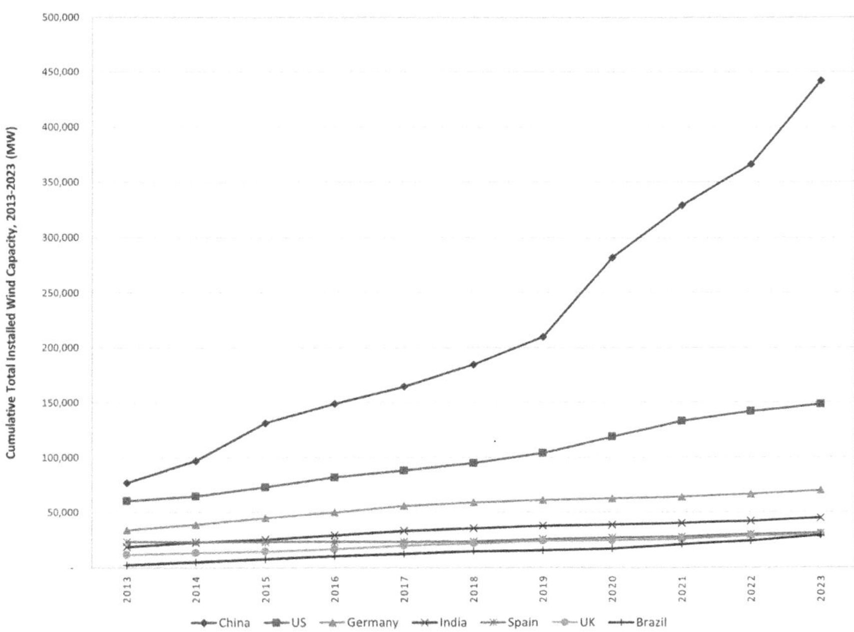

Figure 4.2 Cumulative total installed wind-energy capacity (MW) by top seven countries, 2013–2023. IRENA (2024).

contemporary national priorities, the broader socio-economic context, and their overall economic development goals. Using examples from across the globe, we provide a survey of some of the more widely used national and local policy mechanisms supporting wind-energy production.

4.2.1 Subsidies, Financial and Import–Export Incentives

Subsidies in the energy sector are government interventions directed at influencing the price or cost of the production, distribution, and purchase of energy. Subsidies are commonplace across the energy spectrum and may vary in terms of their size and scope over time. Recent subsidies to the renewable energy sector have drawn some attention by the public, but it is appropriate to recognize that subsidies in the energy sector are long standing and can be important drivers of supply expansions. For instance, in the United States, subsidies to the coal and oil-and-gas industries date back to the early part of the twentieth century. Subsidies are employed by countries across the globe and when appropriately targeted may be put toward reducing the final price paid by the consumer or toward lowering investment, exploration, and production costs.

Financial incentives are critical to sustaining investments in the renewable energy sector. Countries across the world employ such incentives to promote growth and expansion. Incentives and preferential tax structures work in tandem with other policy mechanisms to create an overall situation conducive to the expansion and adoption of wind energy. Incentives may include favorable term, low-interest loans for initial wind-energy investments or joint ventures and technology transfer agreements, financial incentives to local companies that engage in the manufacturing of wind turbines, direct grants to developers, and subsidies on turbine components that need to be imported from elsewhere, among others.

Favorable tax structures may also provide the incentives for growth in this sector. Everything from tax credits, rebates, deductions, favorable depreciation rates, property tax incentives, and exemptions of excise duties to royalties on the adoption of new technologies is a commonly observed strategy (Lewis and Wiser 2007; WindExchange 2025c). Additionally, reductions in value-added and sales taxes on turbine technology, when combined with other mechanisms, may also help with the localization of the technology (Saidur et al. 2010). Studies have long since suggested that long-term consistency in subsidies and other incentive programs is key to the expansion of wind-energy installations (Barradale 2010).

For example, in the United States, the recent Inflation Reduction Act (IRA) of 2022 provides significant production and manufacturing tax credits, financing mechanisms for clean energy investments, technology deployment grants, energy efficiency loans, and opportunities for public–private partnerships to expand clean and renewable energy technologies and programs that have local economic benefits (WindExchange 2025b). The IRA for clean energy projects is slated to phase out in 2032 once greenhouse gas emissions in the energy sector decline to substantially lower levels than seen in 2022. After campaign promises to refocus on oil and gas, it remains to be seen how the 2025 Trump Administration will reshape the renewable wind and solar energy sectors in particular.

India is currently ranked fourth in the world after Germany in installed wind-generating capacity (IRENA 2024). The government of India has refocused efforts to meet a renewable energy target of 500 GW by 2030, with wind energy contributing 140 GW. This target would more than triple the 2024 installed capacity of 44.74 GW (GWEC MEC 2022). To meet these goals, the government has scaled up R&D activities within the sector, encouraged partnerships between entities, focused on incentivizing local turbine development technology, promoted the export of turbines manufactured in India, and continued with previous generation-based incentives for wind farms (Pearson 2013; Shah 2024). Generation-based incentives provide a bonus payment to wind farms per kWh of power contributions to the grid (Mani and Dhingra 2013). If the right policy supports and clear long-term investment signals to wind-energy developers endure, that will create the conditions necessary to assist the country's ambitious goals for the renewable energy sector.

Governments may also influence import–export conditions to favor local industrial sectors. This may include relief from customs or excise duties for imported component parts for turbines and related equipment to help grow the sector, as India has done (Mani and Dhingra 2013). Alternatively, they may provide export credit assistance to help domestic wind developers gain advantages by providing low-interest loans to purchasing parties in export markets (Lewis and Wiser 2007). Another way governments may influence conditions is through connecting overseas development aid to the purchase and distribution of domestic technology as was seen in the case of Germany and Denmark (Lewis and Wiser 2007). Some of these strategies could come under the scrutiny of the World Trade Organization (WTO), if they were seen as violating trade agreements that unfairly advantaged the wind-energy sector of some countries over others.

4.2.2 Feed-in Tariffs

Feed-in tariffs (FiTs) or renewable energy payments are a form of subsidy, and a frequently observed policy instrument targeted at renewable energy producers. They have been adopted across many countries in Europe, Canada, China, and elsewhere to create a demand for wind power. With FiTs, governments require utility distribution companies to buy renewable energy from producers at predetermined prices (Ackermann and Söder 2002; Couture et al. 2010).

Although the specific characteristics of FiTs vary country by country, they have been designed in two principal ways: fixed price tariffs and premium tariffs. Under the fixed price model, governments establish a guaranteed price for renewable energy exceeding the market rate per kilowatt-hour (kWh) for a predetermined period. This mechanism provides long-term stability and reduces investment risk, thereby encouraging new and existing producers to expand their operations (Couture et al. 2010). For example, Germany's early adoption of fixed FiTs contributed significantly to the rapid expansion of its renewable energy sector (Hitaj and Löschel 2019). Denmark similarly initiated early research programs for wind turbines and instituted a FiT for wind energy in 1976 (Johansen 2021). These early subsidies allowed for the establishment of linkages between R&D in turbine manufacturing and energy generation, which propelled them into a leading developer and exporter of wind-energy technologies.

Alternative versions of this policy instrument include the constant or sliding premium FiT, which tags on a premium to the spot market price of electricity (Klein 2008). The premium is intended to approximate the extra cost of generating renewable energy and also account for the positive externalities of renewables. Premiums may, at times, include price caps and price floors to make for a relatively stable investment climate (Couture et al. 2010). Collectively these policies provide stability and greater confidence in future wind-energy markets (Lewis and Wiser 2007). Both fixed and premium FiTs provide a predictable revenue stream, enabling producers to cover high upfront capital and investment costs and achieve a steady, reasonable return on investment.

FiT mechanisms have been shown to encourage the rapid development of renewable energy production by providing long-term fixed price or sliding premium contracts. Agreements may range between 10 and 25 years and are variably structured to encourage technological innovation and decrease the reliance on fixed payments in later years (Klein 2008). Moreover, payments may be graduated by project size and location to meet desired goals of efficiency. FiTs encourage production in wind energy by protecting producers from the volatility of electricity prices in the short run and by providing them with access to the grid. By mitigating these risks, FiTs create an environment conducive to technological innovation and accelerate adoption of new technologies. Examples from countries such as Germany and China provide insights on how governments use tailored FiT mechanisms to balance the goals of encouraging investment and fostering innovation.

Germany has been a global leader in renewable energy adoption, with FiT playing an early and central role. With 24% of its electricity coming from renewable sources by 2012, it set targets of 50% by 2030 and 65% by 2040 (Fulton et al. 2012). Germany's FiT policy emphasizes transparency, longevity, and certainty to incentivize investment. The country employs a tiered approach, where all wind projects start with the same fixed tariff for the initial investment period. After this, tariffs are adjusted based on site-specific wind conditions: higher-resource sites receive reduced tariffs, while lower-resource sites maintain the original rate (Hitaj and Löschel 2019). In 2012, the initial tariff was set at Euro 8.8 cent/kWh with additional bonuses for the use of advanced technologies or replacing older technologies (GWEC 2013).

As wind energy becomes more established and power generation builds, countries adopt a degression rate to progressively reduce tariffs for existing plants or new entrants to the field. Degression rates may be fixed annually, or they can be quite complex in derivation as the sector matures, becomes more competitive, and cost reductions take hold. To reflect the maturing wind sector and decreasing costs, Germany instituted an annual degression rate progressively reducing tariffs for both new and existing projects (IRENA-GWEC 2012).

China's FiT model, introduced in 2009, incorporated geographic differentiation to address the country's diverse wind resource potential. The government categorized the country into four zones based on wind resources, assigning higher tariffs to regions with lower wind potential and lower tariffs to wind-rich zones (Table 4.1; Gipe 2009). This approach ensured balanced development across regions and prevented overconcentration in areas with high wind resources (Li et al. 2014).

The meteoric rise of installed capacity and energy generation established China as the world's leader in the wind-energy sector (see Fig. 4.2). The combined onshore and offshore installed capacity topped 440 GW in 2024 and observed an annual growth rate of 19% between 2013 and 2023 (IRENA 2024). FiT played one

Table 4.1 Renewable energy tariff zones in China, as reported in Gipe (2009).

Tariff zones	Regions of China
Zone 1	Inner Mongolia, excluding Chifeng, Tongliao, Xing-an county, Hulunbei-er; Xinjiang, including Urumqi, Yili-Kazak auto. region, Changji county, Kelamayi, Shi-he-si.
Zone 2	Hebei province, including Zhangjiakou, Chengde; Inner Mongolia, including Chifeng, Tongliao, Xing-an, Hulunbei-er; Gansu province, including Zhangye, Jiayuguan, Jiuquan.
Zone 3	Jilin, including Baicheng, Songyuan; Heilongjiang province, including Jixi, Shuangyashan, Qitaihe, Suihua, Yichun, Daxing-anling; Gansu province, excluding Zhangye, Jiayuguan, Jiuguan; Xinjiang, excluding Urumqi, Yili Kasak auto. Region, Changji, Kelamayi, Shi-he-zi; Ningxia.
Zone 4	Regions not included in zones 1, 2, or 3.

Adapted from Gipe (2009).

part, though there were numerous other governmental supports that aided this rise. Citing high costs along with administrative challenges and increased competitiveness of the renewable energy sector, the Chinese government in 2021 started phasing out FiT in wind energy (Sandalow et al. 2022). They have replaced this with a range of market-driven mechanisms to support wind-energy development. Despite China's remarkable expansion of wind power during the past decade, the country is now focused on addressing obstacles to full grid integration. Expanding transmission line capacity and building system flexibility to accommodate the intermittency of renewable energy supply are some areas of focus (Yu et al. 2024).

FiTs provide a cornerstone policy for advancing renewable energy investments, including wind power. Their ability to keep stable, predictable revenue streams encourages investment and innovation. The design and implementation of FiT must continually be fine-tuned to address market conditions and efficiencies. By learning from successful examples in Germany and China, policymakers could refine FiT to support a resilient wind-energy sector.

4.2.3 Renewable Portfolio Standards

Renewable Portfolio Standards (RPS), known as renewable obligations in the United Kingdom, and sometimes referred to as renewable electricity standards or purchase mandates, are policy instruments designed to increase the share of renewable energy in electricity generation. RPS require electricity suppliers to source a specified proportion of their energy portfolio from renewable resources, such as wind, solar, and geothermal energy. These policies have significantly benefited wind-energy development across the globe.

RPS frameworks often include mechanisms that allow renewable energy producers to trade or sell credits for excess renewable energy production to suppliers unable to meet their targets. This trading system introduces flexibility, encouraging increased production and utilization of renewable energy over time. Additionally, RPS policies frequently mandate a gradual increase in renewable energy targets, driving sustained growth in the sector (USEIA 2012). Non-compliance is typically addressed through penalties imposed on utilities that fail to meet specified standards. The flexible design of RPS policies allows for variation in structure, enforcement, and compliance, reflecting local goals and conditions.

In the United States, 29 states and the District of Columbia have implemented RPS, requiring utility companies to provide a certain percent of overall electricity from renewable or other clean sources (Barbose 2024). In addition to RPS, they use mechanisms such as Clean Energy Standards (CES), Alternative Portfolio Standards (APS), or zero-emission standards, all of which encourage the incorporation of renewables or low or zero carbon-emitting sources.

As a report from the Center for Climate and Energy Solutions suggests, Colorado, North Carolina, and Oregon have set ambitious goals to achieve carbon neutrality or eliminate emissions by mid-century (CCES 2024). Additionally, six states have voluntary goals, while some, like Massachusetts, employ multiple standards. The spillover benefits from RPS can drive renewable energy adoption, reduce greenhouse gas emissions over time, and support job creation in the renewable sector.

Past studies have shown that RPS have buoyed wind-energy investments in such states as Texas, Minnesota, and Iowa, while elsewhere as in Wisconsin, RPS have been responsible for also supporting wind-energy investments in the neighboring state of Iowa (Fig. 4.3; Bird et al. 2005). For instance, in the early 2000s, Iowa's excess capacity in wind production was used to meet its neighboring state, Wisconsin's, RPS mandates. Iowa's early adoption of wind energy and large installed capacity through preferential property and sales taxes contributed to these advantages.

The Australian government has an ambitious target of 82% renewable energy in the Australian grid by 2030. One mechanism is through the Renewable Energy Target (RET) scheme, which aims to increase the share of renewable sources in the generation of electricity each year by 33,000 GWh between 2020 and 2030 (AGCER 2024). Through market incentives it supports both small- and large-scale renewable projects and investments.

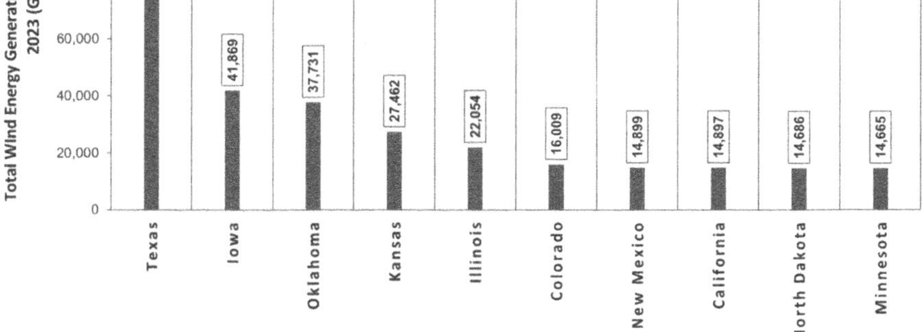

Figure 4.3 Total wind-energy generation in top 10 states in the United States (GWh), 2024. USEIA (2024).

The Renewables Obligation (RO) adopted by the United Kingdom in 2002 mandated electricity suppliers to buy a graduated percentage of their energy from renewable sources. Starting at 3% in 2002–2003, it raised to 10.4% by 2010–2011 and 15.4% by 2014–2016 (Mitchell et al. 2006). RO certificates are issued to regulated renewable energy producers, which then may be traded to meet obligations. Suppliers failing to meet targets must contribute to a buy-out fund, which redistributes excess funds to renewable energy suppliers (Ofgem 2025).

In 2017, the United Kingdom closed the RO scheme to all new generation capacity, though it supports existing generators until phase out (Pashakolaie et al. 2024; Ofgem 2025). The United Kingdom replaced RO with an updated Contracts for Difference (CfD) mechanism that supports low-carbon electricity generation. The mechanism incentivizes and protects producers from price volatility and consumers from increased prices (Watson and Bolton 2024). In 2024, the Labor Government in the United Kingdom vowed to reinvest in wind energy and double production from onshore projects by 2030 (UK Gov 2024).

4.2.4 Local-Content Requirements

Governments may often mandate domestic content requirements in different economic sectors to maximize the benefits accrued from developing specific industries. In the wind-energy sector, countries such as Brazil, Canada, China, and Spain among others have used local-content mandates during the industry's early growth phases. Governments may require component parts to be sourced locally, equipment to be manufactured by domestic companies, or encourage technology transfers through joint ventures. Such measures aim to create jobs, provide skills-based training for local labor, and generate multiplier effects that benefit local economies and tax bases. They are frequently used to nurture infant industries and position them as globally competitive players.

However, local-content requirements can present challenges. The WTO views such measures as potential violations of trade agreements, leading to inefficiencies and increased costs. From a market perspective, they are seen as an inefficient and costly allocation of resources that could drive up prices in the short run and be detrimental to developing competitive industries in the long run. Even so, local-content requirement policies have been relatively common in the wind sector, offering significant short-term benefits.

China's early mandated "Ride the Wind" program was probably one of the more famous local-content requirement policies. Initiated in 1996, it required Chinese-foreign joint ventures in the wind-energy sector to build assembly factories that manufactured a certain percentage of the wind-turbine generator equipment in China itself. The content requirement increased from 50% in 2003 to 70% in 2004, with the policy being eliminated in 2009. The local-content requirement directive has arguably been one factor that led to China's global primacy in wind-power component manufacturing and installation in a short time frame, and provided the country an advantageous playing field in the international wind-energy sector. Installed capacity rapidly increased, the local turbine manufacturing industry benefitted through technology transfers, and it was seen as a significant employment generator (Kuntze and Moerenhout 2013). Although local-content requirement was eliminated in 2009 in favor of FiTs and other measures, its legacy is evident in China's continued dominance in the global wind-energy sector.

Brazil launched its Program of Incentives for Alternative Energy Sources (PROINFA) legislation in 2002. Included in the first phase of PROINFA policies were 60% of local-content requirement (so-called nationalization indices) for wind-turbine equipment and 90% for services (GWEC 2011). In 2002, Brazil had 22 MW of installed wind-power capacity which increased to 931 MW by 2010. By most accounts, this was a modest increase for a country that has significant wind-energy resources (Kuntze and Moerenhout 2013). The stringent local-content requirements dampened the expansion of installations as the pressure on the local steel industry for turbine towers resulted in rising local steel prices (Backwell 2011).

To encourage wind industry investment, the Brazilian government turned to a bidding process and auctions in 2009 to award large-capacity allocations. In early wind-power auction awards, a national procurement stipulation disallowed the import of wind turbines with a nominal capacity less than 1.5 MW. However, the nationalization index of 60% for equipment and parts was only required if subsidized financing was sought from the Brazilian National Development Bank (BNDES). The significantly lower financing costs offered by BNDES resulted in many, but not all, wind projects taking advantage of this subsidy and in turn meeting the 60% nationalization index. Data indicate that this requirement encouraged technology transfer, the rapid expansion of the local manufacturing supply chain, and multiplier effects for regions across Brazil (GWEC 2011).

Spain emerged as a leader in wind energy in Europe, only second after Germany and ranking fifth globally with 30,700 MW of cumulative installed capacity by 2023 (see Fig. 4.1). With a policy mix of FiTs, technology transfers from joint-venture projects, support for key industry players, provincial policies informally mandating local-content requirements, and the financial backing of the national government, Spain's installed capacity of 73 MW in 1994 grew to 22,796 MW by 2012, and over 30,000 MW today (Kuntze and Moerenhout 2013).

Local-content requirements focused on manufacturing turbine parts and equipment were in place in provinces like Galicia, Navarra, and Valencia. Provinces granted development rights only when up to 70% local-content requirements are met, as in the case of Navarra (Hao et al. 2010). While local-content requirement rules are more relaxed today, the early advantages allowed for the expansion of some of the world's largest players in the wind market, including Gamesa (now merged with Siemens) and Acciona Energia (now merged with Nordex) (San Miguel et al. 2010).

Spain's impressive early expansion into wind energy came with a substantial debt burden. Policy shifts over the past decade and more have reigned in large-scale subsidies, including local-content requirements to wind projects (Sills 2010). Based on extensive industry knowledge and experience from onshore wind projects, Spain stands to gain from expanding its significant offshore resources. A regulatory framework outlining a competitive bid process, remuneration, grid capacity, and other features was advanced in 2024 through a new Royal Decree from the Spanish government.

Local-content requirements have faced scrutiny under WTO trade rules, as noted above. In 2013, the WTO ruled against Ontario, Canada, finding its 60% local-content mandate for wind subsidies discriminatory against foreign firms. This led Ontario to revise its policies (WTO 2013). Such disputes highlight the ongoing tension between national industrial policies and international trade agreements as countries seek to balance domestic economic goals with global competition rules.

Local-content requirements in the wind-energy sector have played a pivotal role in fostering domestic manufacturing, creating jobs, and supporting nascent industries. Today, few countries still implement such measures. While there are benefits to be gained early in a sector's development, critics have long since highlighted the inefficiencies that also accrue. The future influence of these policies on the global wind-energy market will depend on their ability to adapt to evolving economic and trade conditions.

4.2.5 Research and Development Investments

Sustained investments in R&D and innovation are critical drivers of the wind-energy industry. Studies suggest that the effectiveness of R&D investments are influenced more by how they are distributed or targeted, as opposed to the total amount of money invested (Lewis and Wiser 2007). Localizing wind-energy technology and expertise offers significant advantages, such as reducing component costs, fostering rapid innovation, and minimizing technology transfer expenses. Additionally, a competitive and expanding domestic manufacturing sector could tap into global export market for wind-turbine components.

Countries adopt diverse R&D strategies tailored to their specific conditions, but these efforts generally involve collaborations between national laboratories, research universities, public agencies, and private-sector partners (Astrand and Neij 2006; ETIPWind 2023). In the United States, the National Renewable Energy Laboratory (NREL) leads renewable energy and energy-efficiency R&D. NREL collaborates with private industries, universities, and governmental agencies to advance wind-energy science, improve technological efficiencies, and facilitate knowledge transfer (NREL 2025). Notable NREL initiatives include research on blade aerodynamics, grid integration, and offshore wind technologies, among other areas.

Similarly, the European Wind Energy Agency, rebranded as WindEurope, today supports the wind industry across Europe and elsewhere with a global network of partners (WindEurope 2025). The European Technology and Innovation Platform on Wind Energy (ETIPWind) established in 2016 is an example of an R&D partnership bringing together researchers, stakeholders, experts, and policy makers to formulate recommendations and strategic plans driving wind-energy growth across the European Union (ETIPWind 2025).

Germany has a robust history of governmental support for wind-energy R&D projects dating back to the 1970s. Grants have been awarded to academic institutions and industries to improve turbine design, enhance testing and evaluation capabilities, and to collect long-term environmental data to inform future expansion projects (Jacobsson and Lauber 2006). For example, the Fraunhofer Institute for Wind Energy Systems (IWES) is funded by various German government ministries and the European Union. A leading applied research institute with clients within Europe and beyond, it engages in research on wind-turbine technology, storage systems, energy management, and grid operations among other aspects of wind-power generation (Fraunhofer IWES 2025).

Denmark is another notable example, with its dedicated wind-energy R&D programs contributing significantly to the sector (Johansen 2021; Elliott et al. 2024). University research programs like Danish Technical University's Wind and Energy Systems department focus on blade optimization, offshore wind development, and digital monitoring systems, among other areas (Fig. 4.4; DTU Wind 2025). Denmark's early research investment efforts also have enabled key companies like Vestas and Ørsted to dominate the global market, showcasing the country's success in translating research into competitiveness.

Research and innovation investments in wind energy are pivotal for reducing costs, enhancing efficiency, and driving global competitiveness. Such investment could foster collaboration among public agencies, research institutions, and private-sector players, accelerating the diffusion of technological advancements to society. Examples from the United States, Germany, Denmark, and elsewhere underscore the importance of innovation in shaping the future of the sector.

4.2.6 Green-Power Pricing Markets

Consumer demand for environmentally friendly energy sources significantly influences policy measures that promote renewable and wind-energy expansion. Green-power pricing has emerged as a viable option, allowing utility companies

Figure 4.4 Test station for large turbines at Høvsøre on the western coast of Jutland, Denmark operated by Risø DTU, as seen in 2013. Meteorological towers record wind, temperature, and air pressure at multiple heights, and these data are compared with performance of several turbine models.

to charge customers a premium to support renewable energy development. These programs may directly fund renewable energy projects, enhance a utility's environmental performance and public image, educate customers about alternative energy sources, and provide consumers with choices to purchase electricity generated through environmentally cleaner methods (Lieberman 2002; IER 2013; USEPA 2025).

Green-power markets are present in various countries, including the United States, European nations, Australia, Japan, and Canada, among others (Bird et al. 2002). These programs operate through different pricing mechanisms, such as voluntary or mandated premiums per kilowatt-hour (kWh) or fixed monthly fees. Utilities that lack renewable energy portfolios due to geographic or other constraints could purchase tradable renewable certificates or green tags from other entities. These certificates represent the environmental benefits of renewable energy generation, such as reduced emissions, and can be sold to consumers willing to pay for their positive environmental externalities (Lieberman 2002; USEPA 2025).

Across the United States, hundreds of utility companies offer green-pricing schemes. Data from the National Renewable Energy Laboratory suggests that in 2023 roughly 9.6 million customers bought 319 terawatt-hours (TWh) in retail renewable electricity sales, representing 8% of all electricity retail sales (O'Shaughnessy 2024). While the total number of customers served by this option is still small, some suggest that the potential for growth is substantial given the green-shift, or the wider acceptance and demand for green electricity certification options. Studies from the United States indicate a higher willingness to pay for green energy when the public is educated about the benefits and the premiums are priced appropriately, suggesting wide potential for these markets (Gustafson et al. 2019; Knapp et al. 2020).

Data from other countries suggest variability in the uptake of green-power pricing markets. Surveys on consumers worldwide still indicate a disconnect between supporting energy transitions and committing to pay premiums for renewable energy. A recent study reported that less than half those surveyed from 10 different countries agreed with a greater than 2% price increase for electricity to reduce greenhouse gas emissions (Dougans et al. 2024). Sweden reported a green-power pricing market of 10.3% in relation to gross electricity consumption back in 2001 (Markard and Truffer 2006). Today it ranks as one of the top three European countries where the share of energy from renewable sources is over 65% (EEA 2025).

Sweden introduced green-power options in 1996 with relatively low premiums, taking advantage of the large proportion of renewable hydropower in its existing electricity portfolio. It also simultaneously started eco-labeling its green electricity products. These along with market advantages pushed Sweden to an early lead position in the green market (Fig. 4.5). More recently, the European Commission

Figure 4.5 Turbines on the southwestern coast of Sweden at Falkenberg, as seen in 2013. Nearly one-fifth of Swedish electric-generating capacity now comes from wind power and nearly half from hydropower. CIA Electricity (2022).

has proposed reforming the electricity market design by providing consumers more options and education on pricing to make informed choices (European Commission 2025).

The overall global impact of green-power pricing markets remains mixed, but it holds potential to support renewable energy development and foster consumer participation, especially in countries yet to see renewables supplying a large proportion of electricity to the consumer. Other policy measures discussed in this chapter likely have a greater influence on renewable energy adoption. As some have suggested customer education and aggressive marketing with certification and labeling programs, offering a range of green-power product options to customers with varying pricing sensitivities, and an open retail market that allows customers to choose utility companies could all help increase consumer participation rates (Bird et al. 2002; Markard and Truffer 2006; Gillenwater et al. 2014).

4.3 Land-Use Decisions and Siting Policies

Local land-use and zoning decisions play important roles in the locational and infrastructure development aspects of both large-scale commercial and smaller-scale individual wind projects. Siting and approval processes are influenced by a complex web of laws and regulations at the local, state, and national levels. These processes often vary significantly across nations, states, provinces, or regions, adding layers of complexity to permitting and development (Lopez et al. 2023; WindExchange 2025d).

As wind energy becomes more prevalent, communities face the challenge of establishing siting guidelines that account for topographic, aesthetic, and environmental factors. For example, wind-project ordinances typically include setback requirements from residences, flicker mitigation strategies, detailed environmental assessments to protect wildlife habitats, noise, and shadow impact studies before project approval. In other cases, local governments might require decommissioning fees to ensure that sites are restored after their operational lifespan.

Larger wind-farm projects require negotiations for land purchases or long-term lease agreements with private landowners, as well as securing rights of way and easements for public lands. Developers also face public attitudes and local opposition, often leading to lawsuits, appeals, and delays in the acquisition and permitting process (Ropeik 2024).

Two Counties, Kansas

Reno and Harvey counties are located next to each other in south-central Kansas. The region is mostly the Arkansas River Lowlands (see Fig. 7.1). This broad and relatively flat terrain is underlain by unconsolidated sand and gravel of the High Plains Aquifer. The wind-energy potential is rated as fair with average wind speed at 50 m height of 6.8 to 7.5 m/s (WindExchange 2008). The issue of developing wind farms in these counties was contentious during the past several years.

NextEra Energy became interested in Reno County in the late 2010 decade in order to take advantage of federal tax credits that would expire in 2020. This set off public opposition based on many perceived environmental and health factors. Citizens were concerned about the conversion of rural aesthetic values for industrial development. The wind farm was blocked by a grass-roots initiative in 2019. The Kansas Appellate Court subsequently ruled in 2022 against NextEra Energy and its Pretty Prairie Wind project (Green 2022), and NextEra cancelled the project (GEMW 2024).

NextEra Energy also explored the possibility for a wind farm in neighboring Harvey County. Once again, public opposition came forward voicing many of the same objections heard in Reno County. Another issue was potential disturbance or contamination of the High Plains Aquifer. In 2022, the county planning commission unanimously approved regulations that included setback distance, noise level, and several other requirements.

The Harvey County Commission then considered the matter. Public opposition continued, nonetheless, and the county commission agreed to a moratorium on wind turbines. The County Commissioners finally voted unanimously in November 2023 to prohibit commercial wind energy in the county. Protection of the High Plains Aquifer was a key issue for this decision. From a geological point of view, however, it is unclear how turbines would have any negative impact on the aquifer.

The City of Halstead, in western Harvey County, announced plans in 2023 with Concurrent for a lease-purchase option to build a battery energy storage systems (BESS), named the Tallgrass Prairie Energy Center. It would store excess wind- and solar-generated electricity from regional sources. The BESS would have a capacity of 200 MW and up to 800 MWh, which is enough electricity to power 200,000 homes for four hours. This is typical for such facilities (Coleman 2025).

(*Continued*)

Two Counties, Kansas (Continued)

(A)

(B)

Groundwater from the High Plains Aquifer in western Harvey County, Kansas. (A) Well house and pumping station for municipal water supply, City of Mound Ridge. (B) high-capacity pump (right) feeds water into a center-pivot irrigation system (left).

Early in 2025, Halstead's Planning and Zoning Commission released its proposed BESS regulations. The regulations cover all aspects of construction, operation, safety, emergency procedures, and eventual decommissioning of the facility. A bill under consideration in the Kansas Legislature would exempt BESS facilities from property taxes for 10 years (Kansas Legislature 2025). Meanwhile, Concurrent offered to pay the city $500,000 per year for placing the BESS in the industrial park as well as to purchase all necessary equipment for the fire department (Bush 2025b).

As with the wind farms, public opposition appeared against the BESS, in spite of its planned location in an existing industrial park. The primary objection apparently was its risk of fire or explosion. After a thorough review of the issues, the city Fire Chief noted that other industries already present in

Two Counties, Kansas (Continued)

Yard sign opposed to construction of a BESS in the Halstead industrial park, Kansas.

the city represent greater fire or explosion risks, and he was mostly comfortable with a BESS in the industrial park (Bush 2025a). At the time of book writing, both Harvey County and the City of Halstead had approved temporary moratoriums for BESS until January 2028.

Siting and permitting requirements vary based on the size and scale of projects. Communities have developed ordinances addressing various considerations associated with wind farms. A recent analysis concluded that across the United States, communities have developed more than 1,800 local zoning ordinances for wind energy (WindExchange 2025a). The ongoing shift from fossil fuels to renewable electricity generation often pits urban-industrial demand against rural-agricultural land use, as large amounts of land are necessary for wind and solar facilities (Coleman 2025). The so-called *agrivoltaic* projects are designed so the land may continue for livestock grazing, growing crops, and pollinator habitats.

Legal aspects for siting wind farms may be quite complex in some situations. In Oklahoma, for example, Enel constructed a 150-MW wind farm with 84 turbines on the Osage Mineral Estate. The company had secured surface rights, but did not obtain mineral rights. The dispute began in 2014, and finally a federal judge ruled in 2024 that Enel had violated federal and tribal laws regarding the use of the mineral estate (Soler 2024). Enel was ordered to remove the wind farm and pay damages and attorney fees.

The following list provides some of the general ordinance considerations observed across communities within the United States (WindExchange 2025d). Similar factors apply in many other countries.

- Turbine height regulations—Defined from the base to the blade tip at its highest point.
- Tower spacing requirements—Minimum distances between turbines within a facility.
- Setback regulations—Distances from roads, power lines, adjacent properties, land uses, or inhabited structures.
- Minimum acreage—Ensures adequate spacing and safety for the number of turbines.
- Proximity restrictions—Distance from schools, airstrips, and residential areas.
- Noise standards—Limits on maximum allowable noise levels from operations.
- Flicker regulations—Mitigation of shadow flicker impacts on adjacent properties and occupied structures.
- Wildlife and avian considerations—Measures to minimize harm to local wildlife and ecosystems.
- Environmental protections—Safeguards for wetlands, water bodies, and other sensitive areas.
- Appearance standards—Non-reflective and neutral turbine colors to reduce visual impact.
- Decommissioning and restoration—Ensures proper site rehabilitation and turbine removal post-operation (see sect. 3.3).
- Zoning regulations—Defines permissible wind-farm locations within a community.
- Signal interference—Prevents disruptions to television, radio, and other communication signals.

Wind farms generate sizable revenues for communities in terms of lease and royalty payments, property taxes, allied business, and new construction and service jobs. Despite these advantages, wind projects may not garner the unanimous approval of the communities in which they locate. Success in siting and permitting depends on comprehensive land-use planning, systematic evaluation of proposals, and active engagement with public and stakeholder communities. Communities that integrate these practices are better positioned to address concerns and harness the economic and environmental benefits of wind energy.

4.4 Summary

This chapter provides a survey of some often-used policy instruments that guide development in the wind-energy sector. Using examples from across the globe, the chapter outlines how countries adopt measures to spur growth in this industry and generate multiple benefits across communities.

Countries rarely rely on a single policy instrument, instead employing a combination of mechanisms tailored to their socio-economic contexts. Thus, as in the case of India, we see the government adopting not only a host of subsidies and tax incentives, but also depreciation bonuses and relief from excise and customs duties. China supported its early wind-energy sector through reliance on local-content requirements, then shifting to FiTs, while also supporting R&D and other subsidies. The United States relies on a system of RPS and significant investments in R&D into wind-energy technology. These strategies reflect the diverse approaches taken to meet specific national and regional goals while addressing broader socio-economic priorities.

Ensuring that policy instruments achieve their intended objectives is critical to accountability and effective public policy. Based on the work of Enzensberger et al. (2002, p. 796) and others, a set of six criteria are used to evaluate policy instruments. Each of these provides a way to gauge the impact of policy mechanisms on their intended targets. These include:

- Effectiveness—Measures the extent to which goals and targets are achieved, along with the time required to meet them.
- Efficiency—Assesses resource allocation, emphasizing the highest outcomes relative to capital investments.
- System conformity—Examines how well an instrument aligns with existing market and legal frameworks.
- Practicability—Considers the ease of implementation and regulation.
- Flexibility—Evaluates adaptability to future conditions, including regional variations and additional instruments.
- Economic impact—Gauges how policy instruments influence regional economic development.

While policy instruments are essential, broader enabling conditions also play a pivotal role in the successful expansion of wind energy. These include political commitment and governmental support to creating a conducive environment for investment and innovation, transparent and streamlined administrative and permitting processes, clear pricing structures, reliable grid access and connectivity, financing mechanisms, and public commitment achieved through stakeholder engagement and awareness campaigns, all of which are important components to a successful wind-energy sector. Countries that continually fine-tune effective policy frameworks with these enabling factors are better positioned to realize the economic, environmental, and societal benefits of wind energy.

References

Ackermann, T. and Söder, L. 2002. An overview of wind energy-status. *Renewable and Sustainable Energy Reviews* 6(1–2), p. 67–127.

AGCER 2024. Renewable energy target. Australian Government Clean Energy Regulator. https://cer.gov.au/schemes/renewable-energy-target

Astrand, K. and Neij, L. 2006. An assessment of governmental wind power programmes in Sweden using a systems approach. *Energy Policy* 34, p. 277–296.

Backwell, B. 2011. In depth: Wind industry is caught in a Brazilian steel-price trap. Recharge. https://www.rechargenews.com/wind/in-depth-wind-industry-is-caught-in-a-brazilian-steel-price-trap/1-1-842708

Barbose, G. 2024. U.S. State renewables portfolio and clean electricity standards: 2024 status update. Lawrence Berkley National Laboratory: Energy Markets and Policy Report. https://eta-publications.lbl.gov/sites/default/files/lbnl_rps_ces_status_report_2024_edition.pdf

Barradale, M.J. 2010. Impact of public policy uncertainty on renewable energy investment: Wind power and the production tax credit. *Energy Policy* 38(12), p. 7698–7709.

Bird, L., Bolinger, M., Gagliano, T., Wiser, R., Brown, M. and Parsons, B. 2005. Policies and market factors driving wind power development in the United States. *Energy Policy* 33, p. 1397–1407.

Bird, L., Wüstenhagen, R. and Aabakken, J. 2002. A review of international green power markets: Recent experience, trends, and market drivers. *Renewable and Sustainable Energy Reviews* 6, p. 513–536.

Bush, B. 2025a. Halstead Fire Chief shares his thoughts about battery storage. *Harvey County Now* 10(34), p. 1–8.

Bush, B. 2025b. Mayor: Halstead may have a lot to gain from a BESS. *Harvey County Now* 10(29), p. 1–s.

CCES 2024. U.S. State electricity portfolio standards. Center for Climate and Energy Solutions. https://www.c2es.org/document/renewable-and-alternate-energy-portfolio-standards/

CIA Electricity 2022. Electricity generation sources. U.S. Central Intelligence Agency. https://www.cia.gov/the-world-factbook/field/electricity-generation-sources/

Coleman, K. 2025. Energy gap: Renewable paired with batteries. *Colorado Country Life* 56(1), p. 8–9.

Couture, T., Cory, K., Kreycik, C. and Williams, E. 2010. A policymaker's guide to feed-in tariff policy design. Technical Report NREL/TP-6A2-44849, National Renewable Energy Laboratory, U.S. Dept. of Energy, p. 144. https://www.nrel.gov/docs/fy10osti/44849.pdf

Dougans, G., Scalise, J., Farinha, A., Muscat, O. and Sapre, A. 2024. Will vs. wallet: Consumers want green energy, but not at any cost. Bain & Company Brief.

https://www.bain.com/insights/will-versus-wallet-consumers-want-green-energy-but-not-at-any-cost/

DTU Wind 2025. Department of Wind and Energy Systems. https://wind.dtu.dk/

EEA 2025. Share of energy consumption from renewable sources in Europe. European Environment Agency. https://www.eea.europa.eu/en/analysis/indicators/share-of-energy-consumption-from

Elliott, C., Schumer, C., Gasper, R., Ross, K. and Singh, N. 2024. A sustained portfolio of policies have transformed Denmark's power sector. World Resources Institute. https://www.wri.org/update/sustained-portfolio-policies-have-transformed-denmarks-power-sector

Enzensberger, N., Wietschel, M. and Rentz, O. 2002. Policy instruments fostering wind energy projects – A multi-perspective evaluation approach. *Energy Policy* 30(9), p. 793–801.

ETIPWind 2023. Strategic research & innovation agenda 2025-2027. https://etipwind.eu/files/file/agendas/231205-ETIPWind-SRIA.pdf

ETIPWind 2025. European technology and innovation platform for wind energy. ETIPWind. https://etipwind.eu/

European Commission 2025. Protecting and empowering energy consumers. https://energy.ec.europa.eu/topics/markets-and-consumers/energy-consumers-and-prosumers/protecting-and-empowering-energy-consumers_en

Fraunhofer IWES 2025. Fraunhofer institute for wind energy systems. https://www.iwes.fraunhofer.de/en.html

Fulton, M., Capalino, R. and Auer, J. 2012. The German feed-in-tariff: Recent policy changes. Deutsche Bank Group: DB Climate Change Advisors, Technical Report. New York. https://www.longfinance.net/documents/1038/dbcca_germanfit_2012.pdf

GEMW 2024. Pretty Prairie wind farm. Global Energy Monitor Wiki. https://www.gem.wiki/Pretty_Prairie_wind_farm

Gillenwater, M., Lu, X. and Fischlein, M. 2014. Additionality of wind energy investments in the U.S. voluntary green power market. *Renewable Energy* 63, p. 452–457.

Gipe, P. 2009. China launches differentiated wind energy tariffs. https://grist.org/article/2009-09-01-china-launches-differentiated-wind-energy-tariffs/

Green, J. 2022. Appeals court rules against NextEra Energy in a lawsuit against Reno County over wind farm. The Hutchinson News. https://www.hutchnews.com/story/news/2022/08/29/court-rules-reno-county-nextera-lawsuit-over-wind-farm-pretty-prairie/7929811001/

Gustafson, A., Goldberg, M., Rosenthal, S., Kotcher, J., Maibach, E. and Leiserowitz, A. 2019. Who is willing to pay more for renewable energy? Yale University and George Mason University. Yale Program on Climate Change Communication, New Haven, CT. https://climatecommunication.yale.edu/publications/who-is-willing-to-pay-more-for-renewable-energy/

GWEC 2011. Analysis of the regulatory framework for wind power generation in Brazil, Summary Report. Global Wind Energy Council, Belgium, p. 48.

GWEC. 2013. Global wind report: Annual market update, 2012. Belgium, Global Wind Energy Council, p. 72.

GWEC MEC 2022. India wind energy market outlook 2022–2026. Global Wind Energy Council & MEC. https://www.gwec.net/reports/indiaoutlook/2022

Hao, M., Mackenzie, M., Pomerant, A. and Strachran, K. 2010. Local content requirements in British Columbia's wind power industry. Victoria, BC, Canada, Pacific Institute for Climate Solutions, p. 26.

Hitaj, C. and Löschel, A. 2019. The impact of a feed-in tariff on wind power development in Germany. *Resource and Energy Economics* 57, p. 18–35. ISSN 0928-7655. https://doi.org/10.1016/j.reseneeco.2018.12.001

IER 2013. Evaluating voluntary consumer adoption of green pricing programs. Institute for Energy Research, White Paper, p. 8. https://www.instituteforenergyresearch.org/wp-content/uploads/2013/06/Green-Pricing-White-Paper.pdf

IRENA 2024. Renewable energy statistics 2024. International Renewable Energy Agency, Abu Dhabi. ISBN ISBN: 978-92-9260-614-5. https://www.irena.org/-/media/Files/IRENA/Agency/Publication/2024/Jul/IRENA_Renewable_Energy_Statistics_2024.pdf

IRENA-GWEC 2012. 30 years of policies for wind energy: Lessons from 12 wind energy markets. International Renewable Energy Agency and Global Wind Energy Council, Abu Dhabi. https://www.irena.org/-/media/Files/IRENA/Agency/Publication/2013/GWEC_WindReport_All_web-display.pdf

Jacobsson, S. and Lauber, V. 2006. The politics and policy of energy system transformation: Explaining the German diffusion of renewable energy technology. *Energy Policy* 34(3), p. 256–276.

Johansen, K. 2021. Blowing in the wind: A brief history of wind energy and wind power technologies in Denmark. *Energy Policy* 152, p. 112139. ISSN ISSN 0301-4215. https://doi.org/10.1016/j.enpol.2021.112139

Kansas Legislature 2025. Bill and laws: HB 2083. Kansas Legislative Sessions. https://kslegislature.gov/li/b2025_26/measures/hb2083/

Klein, A. 2008. Feed-in-tariff designs: Options to support electricity generation from renewable energy sources. VDM Verlage De. Muller Aktiengesellschaft & Co, Saarbrucken, Germany.

Knapp, L., O'Shaughnessy, E., Heeter, J., Mills, S. and DeCicco, J. 2020. Will consumers really pay for green electricity? Comparing stated and revealed preferences for residential programs in the United States. *Energy Research & Social Science* 65, 101457. https://doi.org/10.1016/j.erss.2020.101457

Kuntze, J.-C. and Moerenhout, T. 2013. Local content requirements and the renewable energy industry – A good match? International Centre for Trade and

Sustainable Development, Geneva, Switzerland, p. 67. https://unctad.org/system/files/non-official-document/DITC_TED_13062013_Study_ICTSD.pdf

Lewis, J.I. and Wiser, R.H. 2007. Fostering a renewable energy technology industry: An international comparison of wind industry policy support mechanisms. *Energy Policy* 35(3), p. 1844–1857.

Li, C., Li, P. and Feng, X. 2014. Analysis of wind power generation operation management risk in China. *Renewable Energy* 64, p. 266–275.

Lieberman, D. 2002. Green pricing at public utilities: A how-to guide based on lessons learned to date. Center for Resource Solutions, San Francisco, CA, p. 49.

Lopez, A., Cole, W., Sergi, B., Levin, A., Carey, J., Mangan, C., Mai, T., Williams, T., Pinchuk, P. and Gu, J. 2023. Impact of siting ordinances on land availability for wind and solar development. *Nature Energy* 8, p. 1034–1043. https://doi.org/10.1038/s41560-023-01319-3

Mani, S. and Dhingra, T. 2013. Policies to accelerate the growth of offshore wind energy sector in India. *Renewable and Sustainable Energy Reviews* 24, p. 473–482.

Markard, J. and Truffer, B. 2006. The promotional impacts of green power products on renewable energy sources: Direct and indirect eco-effects. *Energy Policy* 34, p. 306–321.

Mitchell, C., Bauknecht, D. and Connor, P.M. 2006. Effectiveness through risk reduction: A comparison of the renewable obligation in England and Wales and the feed-in system in Germany. *Energy Policy* 34, p. 297–305.

NREL 2025. Wind research. National Renewable Energy Laboratory. http://www.nrel.gov/wind/

Ofgem 2025. Renewables obligation (RO). https://www.ofgem.gov.uk/environmental-and-social-schemes/renewables-obligation-ro

O'Shaughnessy, E. 2024. The state of the U.S. voluntary green power market (2023 Data). National Renewable Energy Laboratory. Renewable Energy Markets Conference Denver, CO. https://www.nrel.gov/analysis/assets/pdfs/status-and-trends-2023-data.pdf

Pashakolaie, V., Cotton, M. and Jansen, M. 2024. The co-benefits of offshore wind under the UK renewable obligation scheme: Integrating sustainability in energy policy evaluation. *Energy Policy* 192, 114259. https://doi.org/10.1016/j.enpol.2024.114259

Pearson, N.O. 2013. India cabinet approves wind farm subsidy. Bloomberg News, August 12, 2013. http://www.bloomberg.com/news/2013-08-13/india-cabinet-approves-wind-farm-subsidy-official-says.html

Ropeik, A. 2024. The siting of an offshore wind port raises new conflicts in Maine. Grist, July 15, 2024. https://grist.org/energy/offshore-wind-port-siting-raises-new-conflicts-for-coastal-mainers-environmental-activists/

Saidur, R., Islam, M.R., Rahim, N.A. and Solangi, K.H. 2010. A review on global wind energy policy. *Renewable and Sustainable Energy Reviews* 14, p. 1744–1762.

San Miguel, G., del Rio, P. and Hernández, F. 2010. An update of Spanish renewable energy policy and achievements in a low carbon context. *Journal of Renewable and Sustainable Energy* 2, p. 031007.

Sandalow, D., Meidan, M., Andrews-Speed, P., Hove, A., Qui, S. and Downie, E. 2022. Guide to Chinese climate policy 2022. Oxford Institute of Energy Studies, Oxford. https://chineseclimatepolicy.oxfordenergy.org/

Shah, R. 2024. Redirecting wind energy in India. Ember. https://ember-energy.org/latest-insights/redirecting-wind-in-india/

Sills, B. 2010. Spain said to save $1.5 billion with cuts for wind, solar & thermal power. Bloomberg News, July 9, 2014. http://www.bloomberg.com/news/2010-07-09/spain-said-to-save-1-5-billion-with-cuts-for-wind-solar-thermal-power.html

Soler, D.M. 2024. Enel ordered to tear down wind farm in the U.S. and pay damages. Notebookcheck, News, December 27, 2024. https://www.notebookcheck.net/Enel-ordered-to-tear-down-wind-farm-in-the-U-S-and-pay-damages.937644.0.html

UK Gov 2024. Policy statement on onshore wind. UK GOV: Policy Paper. https://www.gov.uk/government/publications/policy-statement-on-onshore-wind/policy-statement-on-onshore-wind

USEIA 2012. Most states have renewable portfolio standards. U.S. Energy Information Administration. http://www.eia.gov/todayinenergy/detail.cfm?id=4850

USEIA 2023. Wind explained: History of wind power. U.S. Energy Information Administration. https://www.eia.gov/energyexplained/wind/history-of-wind-power.php

USEIA 2024. Electricity data browser. U.S. Energy Information Administration. https://www.eia.gov/electricity/data/browser/

USEPA 2025. U.S. electricity grid and markets. U.S. Environmental Protection Agency. https://www.epa.gov/green-power-markets/us-electricity-grid-markets

Watson, N. and Bolton, P. 2024. Contracts for difference. House of Commons Library Research Briefing #9871. House of Commons Library. https://researchbriefings.files.parliament.uk/documents/CBP-9871/CBP-9871.pdf

WindEurope 2025. Homepage. https://windeurope.org/

WindExchange 2008. Kansas 50-meter community-scale wind resource map. WINDExchange, U.S. Department of Energy. https://windexchange.energy.gov/maps-data/44

WindExchange 2025a. Permitting and zoning. Office of Energy Efficiency & Renewable Energy. https://windexchange.energy.gov/projects/permitting

WindExchange 2025b. Wind energy financial incentives. Office of Energy Efficiency & Renewable Energy. https://windexchange.energy.gov/projects/incentives

WindExchange 2025c. Wind exchange. Office of Energy Efficiency & Renewable Energy. https://windexchange.energy.gov/

WindExchange 2025d. Wind project site selection. Office of Energy Efficiency & Renewable Energy. https://windexchange.energy.gov/projects/site-selection

WTO 2013. Dispute settlement DS412: Canada – Certain measures affecting the renewable energy generation sector. World Trade Organization. https://www.wto.org/english/tratop_e/dispu_e/cases_e/ds412_e.htm

Yu, A., Lu, S., O'Malia, K. and Prasad, S. 2024. China continues to lead the world in wind and solar, with twice as much capacity under construction as the rest of the world combined. Global Energy Monitor Report. https://globalenergymonitor.org/report/china-continues

5

Environmental and Aesthetic Issues

Wind power has become an essential aspect of fossil-carbon abatement for global environmental benefit (Twidell 2009, p. 26).

5.1 Introduction and Risks

Construction and operation of wind farms containing dozens to hundreds of large turbines represent changes in land use that may have environmental and health consequences and affect the aesthetic appearance of the windscape. In some cases, these impacts are minimal, short lived, or may be mitigated in various ways. In other instances, wind farms have introduced significant disruptions in natural or cultural conditions. Clearly wind farms are not appropriate in some settings, such as urban areas, national and provincial parks, world heritage sites, wildlife refuges, or other places with great natural or cultural values.

On the other hand, extensive areas of agricultural and grazing lands as well as shallow offshore locations are suitable in general and should be considered for potential development of wind energy (Fig. 5.1). Wind farms are, in general, compatible with rural, agricultural rangeland and cropland, including both dryland and irrigated farming. Thus, most wind farms have been sited in pre-existing cropland or pasture locations. Cropland typically has little habitat value, but pastures that are native prairie may provide valuable wildlife habitat (Fig. 5.2).

Growth of wind energy worldwide represents an equivalent non-development of coal mines, oil and gas fields, hydroelectric projects, and nuclear energy. These energy sources often have serious environmental and cultural impacts as well as safety risks that, in some cases, could be far greater than the consequences and hazards associated with wind energy. Earthquake and tsunami destruction of the nuclear power plant at

Wind Energy Landscape: Principles, Techniques, History, First Edition. James S. Aber, Susan E. W. Aber, and Firooza Pavri.

Published 2026 by John Wiley & Sons, Inc.

Figure 5.1 Wild daisies and clover bloom in this hay field within the Prairie Queen wind farm built in 2019 in southeastern Kansas. Vast areas in the Great Plains of North America are available for wind-power development with minimal environmental impact.

Figure 5.2 These turbines stand in a field of tallgrass prairie that provides wildlife habitat. The Diamond Vista wind farm was constructed in 2018 on the Missouri–Arkansas drainage divide in east-central Kansas.

Fukushima, Japan, in 2011 and ongoing collapse of coal mines around the world are reminders of the risks associated with these forms of energy.

Likewise, natural gas explosions continue to take a routine toll, which results in human injuries and deaths as well as substantial property damages. Transporting oil via trains also has led to many accidents, among the most serious the 2013 tragedy at Lac-Megantic, Quebec, Canada, in which 47 people died (Tinker 2014). Among the fastest growing sources of oil is tar sand of Alberta, Canada. Production is more than three million barrels of oil per day, which contributes to Canada being the world's fourth largest oil producer and the leading exporter of oil into the United States. Many people consider this enterprise dubious from an energy economic standpoint as well as an environmental disaster in the making (Kusnetz 2021).

As these examples demonstrate, all forms of energy extraction, production, and delivery carry substantial environmental consequences and human risks. In this context, wind energy is relatively safe; although, like any human enterprise, wind power is not totally free from risks. Accidents, injuries, and rarely deaths have taken place in connection with wind turbines and farms. Most are types associated with many kinds of construction, transportation, and industrial activities.

Wind energy likewise generally has limited environmental impacts. In the offshore setting, for example, early Danish research demonstrated that *offshore wind farms, if placed right, can be engineered and operated without significant damage to the marine environment and vulnerable species,* and that *under the right conditions, even big wind farms pose low risks to birds, mammals and fish, even though there will be changes in the living conditions of some species by an increase in habitat heterogeneity* (Danish Offshore Wind 2006).

Dogger Bank, North Sea

Dogger Bank is a large, shallow, sand bank that covers approximately 18,000 km^2 in the middle of the southern North Sea. The bank extends generally from southwest to northeast, some 260 km long, with the largest western portion in the United Kingdom Exclusive Economic Zone (EEZ). The smaller eastern end includes Dutch, German, and Danish EEZ sectors. The bank is historically well known for fishing, particularly cod, haddock, herring, and other species. Average depth of the bank is about 30 m, which makes it suitable for constructing wind-turbine foundations on the seafloor.

Dogger Bank is generally regarded as a large glacial moraine composed of layered and compacted sediments. However, during preliminary geological investigations some zones were found to have chaotic underground structures. Further detailed study revealed that Dogger Bank was formed by repeated episodes of ice-sheet deposition and deformation that built up a large ice-shoved ridge complex during the last major glaciation, some 15,000 to 25,000 years ago (Phillips et al. 2017). As a consequence, the geological structure beneath

(*Continued*)

Dogger Bank, North Sea (Continued)

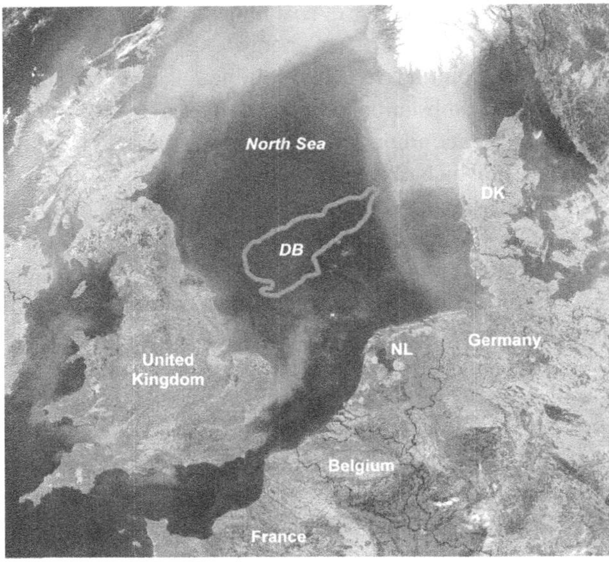

Location of Dogger Bank (DB – outline) in the southern North Sea and surroundings. NASA/Wikimedia Commons/Public Domain.

Dogger Bank is locally riddled with large faults, folds, and other disturbances. These structures extend 40–50 m below the seafloor and render the subsurface unsuitable for wind-turbine foundations. The solution for construction of wind turbines is simply to avoid such zones of known glacial deformation.

Four large wind farms have been authorized in the U.K. sector of Dogger Bank. In addition, a Belgian, Dutch, German, and Danish consortium is planning to build central hubs for connecting power sources and exchanging electricity among the North Sea countries. This approach would utilize the hubs-and-spokes concept of artificial islands and submarine cables connecting nearby wind farms at the northeastern end of the bank. The goal of the consortium is to install 65 GW of offshore wind power by 2030 with expansion to 150 GW by 2050 (NSWPH 2022).

The environmental issues concerning such marine construction come mainly during the building phase due to changes or removal of habitats as well as acoustic disturbance. Marine mammals, fish, and seabirds all would be impacted. In general, Dogger Bank does not host high densities of fish, marine mammals, and seabirds, but harbor porpoise (*Phocoena phocoena*) is an exception in the western portion of the bank (DHI 2018). The northeastern portion of the bank, on the other hand, contains the least potential ecological issues for construction of energy hubs and related wind farms. The U.K. sector of Dogger Bank is a Marine Protected Area (MPA). A ban on bottom-towed fishing came into force in 2022, and since then such fishing has nearly ended (Oceanographic Staff 2022). This dramatic result shows that sensitive MPAs may be protected and, thus, allowed to recover.

The greatest wildlife hazard posed by wind turbines is for flying animals, namely birds and bats. Ground-based wildlife is generally less affected by turbines and wind farms. In fact, deer and other animals may take advantage of turbine shadows to find a spot of relief from the sun during hot summer weather in some Kansas wind farms. Bird mortality at wind farms emerged as a serious issue during the California wind boom of the 1980s, but bat risk did not become apparent until early in the twenty-first century. In both cases, a great deal of research has been conducted, and means are now available to reduce mortality at turbines and to mitigate wildlife issues. Among the negative aspects of utility-scale wind farms, wildlife, visibility, and noise are significant issues. Each of these is examined in more detail in the following sections.

5.2 Wildlife—Birds

In the United States, wildlife is the property of all citizens and, thus, is protected by various federal and state governments. In particular, all species of raptors are protected under U.S. federal and state regulations (Fig. 5.3). Relevant laws include the Migratory Bird Treaty Act of 1918, the Bald Eagle Act of 1940, which also protects golden eagles, and the Endangered Species Act of 1973. According to the U.S. Bureau of Land Management, *it is illegal to capture or kill a raptor; possess a raptor (living or dead), or any pieces or parts of raptors, including feathers, without proper permits from state and federal government agencies* (BLM 2024, p. 12).

Figure 5.3 Bald eagle (*Haliaeetus leucocephalus*) in flight over Lake Kahola in east-central Kansas. Bald eagles have made a remarkable recovery during the past several decades.

Altamont Pass, California

The wind-energy complex at Altamont Pass is among the oldest and largest collections of wind turbines in the United States and is most noted for high bird mortality (see Fig. 1.16). Early research on avian mortality revealed that many birds of prey (raptors) were killed there, which ignited controversy between proponents of wind energy and environmentalists (Lowitz 2019). In the early 2000s, more than a thousand birds of all kinds were killed each year, approximately half of which were raptors including golden eagle, red-tailed hawk, and burrowing owl, as well as kestrels, falcons, vultures, and owl species (CBD 2004). The Altamont Pass vicinity has, in fact, the densest nesting population of golden eagles (*Aquila chrysaetos*) in the world (Lynes 2023).

Field studies indicated an overall mortality rate of roughly one bird death per year for each five turbines (Thelander et al. 2003). Several factors emerged as most likely to produce raptor fatalities, namely ridge-top location, strings of turbines, larger rotor diameter, faster rotor-tip speed, total rotor-swept area within strings, and pocket gopher burrows. However, the type of tower, namely truss or tubular, appeared not to be a significant factor for avian risk. In the first decade of this century, Altamont Pass had some 5,000 turbines that were relatively old and small in size and capacity. According to Lowitz (2019), *the sheer number of turbines in Altamont Pass is a major reason for the high number of bird fatalities in the area.* Newer wind-energy facilities, in contrast, typically have far fewer, taller, and larger turbines that are spaced farther apart, rotate at slower rates, and have much larger generating capacity.

Tight clustering of 1980s vintage U.S. Windpower turbines near a ridge top in the Altamont Pass vicinity, as seen in 2012. Most such turbines have now been removed. See also Fig. 1.15.

Altamont Pass, California (Continued)

The bird-kill fiasco at Altamont Pass came to a head early in the twenty-first century, when the Center for Biological Diversity blamed federal and state agencies for failure to take regulatory action and cited the wind-power industry for ignoring the high levels of bird casualties. The California Energy Commission recommended several steps to reduce and mitigate bird fatalities (CBD 2004). Among the most important steps was repowering—replacing thousands of old, small wind turbines with fewer, larger, modern turbines. These procedures began to yield quick results. Between 2005 and 2010, estimated bird deaths at Altamont Pass declined by half; however, the problem of bird mortality persisted. Eventually Audubon Society filed a lawsuit in 2021 against Alameda County for failing to adopt recommendations of the U.S. Fish & Wildlife Service and the California Department of Fish & Wildlife. The lawsuit was settled in 2023 with plans to minimize and mitigate impacts to eagles and other birds for a new wind-power facility (Lynes 2023).

Bird collisions and mortality, especially for eagles, emerged as a problem in central Wyoming following construction of wind farms near Glenrock in 2008–2010. Two solutions have been implemented, one low-tech method and one high-tech approach. At the Glenrock and Rolling Hills wind farms, 36 turbines have one blade painted black as a test of reducing collision risk for eagles, other birds, and bats. This effort is supported by a consortium of governmental agencies, Audubon, Oregon State University, and energy companies. The U.S. Federal Aviation Administration (FAA) gave special approval for this color scheme; the project began in 2024, and it will continue for several years to evaluate the effectiveness of this simple and low-cost technique (Fig. 5.4).

According to Audubon (George 2023), the black-blade experiment *is guided by the theory that contrast painting reduces motion smear, the visual effect where fast-moving objects appear blurry*. The idea is that birds and bats could perceive fast-moving blades better and, thus, reduce collisions significantly. A similar experiment in Norway showed greater than 70% reduction of mortality for white-tailed eagles (*Haliaeetus albicilla*). Black-and-white stripes have been used likewise on highways for more than a decade to improve lane visibility for drivers at high speed (NDOT 2013).

Nearby at the Top of the World wind farm, the IdentiFlight system utilizes a network of cameras to monitor birds flying within the wind farm. AI looks in particular for golden eagles (*Aquila chrysaetos*) and bald eagles (*Haliaeetus leucocephalus*), and when these birds are spotted nearby turbines are shut off. Top of the World is named for a ridge where golden eagles roost, and some eagles were killed

Figure 5.4 Rolling Hills wind farm. (Left) GE Wind 1.5 MW turbines with one blade painted black (2024). Hub height 80 m; rotor diameter 77 m; year online 2009. Hoen et al. (2018–2025). (Right) close-up view of single blade; outer two-thirds are black (front and back). Elevation about 1,650 m; near Glenrock, Wyoming.

Figure 5.5 Top of the World wind farm went online in 2010 with 110 GE Wind 1.5 MW turbines. Hub height 80 m; rotor diameter 77 m. Hoen et al. (2018–2025). The clustering of multiple turbines poses an obvious risk for flying animals—eagles, other birds, and bats. Elevation about 1650 m; near Glenrock, Wyoming.

by turbine blades in violation of the Migratory Bird Treaty Act (Fig. 5.5). Field testing of the IdentiFlight technology began in 2014, and now the entire wind farm of 110 turbines is protected by monitoring. Each detection unit consists of eight cameras mounted on a tower. The wide-angle cameras span the entire horizon, and when an eagle is spotted another robotic camera tracks its flight. An independent study showed in 2020 that eagle deaths had declined by more than 80% (Wells 2021).

Across the United States, except for California and Wyoming, wind power is generally not responsible for a significant number of bird deaths (GAO 2005). The U.S. Fish & Wildlife Service estimated as of 2017 that on average every year at least seven million birds are killed by collisions with communication towers (Fig. 5.6),

Figure 5.6 Communication towers kill several millions of birds every year in the United States. Typical examples from the Flint Hills of east-central Kansas.

25 million with electrical lines (see Fig. 3.18), 215 million with vehicles, and 600 million die from colliding with building glass. Cats kill 2.4 billion birds per year. In contrast, wind turbines account for some 234,000 bird deaths annually (USFWS 2017). These numbers put into context the relatively minor role of wind turbines for bird mortality in the United States.

A similar situation exists in Europe. According to WindEurope (2024), *the impact of wind farms on birds and bats is extremely low compared to the impact of climate change and other human activity*. Furthermore, *research shows 0.3 bird fatality per gigawatt-hour of wind compared with 5.2 bird fatalities per gigawatt-hour of fossil fuel generation*. Still, some sources claim that wind turbines kill tens of millions of birds annually worldwide (Wiegand 2013). Such claims are not supported by biological research and serve to polarize public perception over the issue of wind-turbine risk for birds. On the other hand, several large environmental organizations, such as Greenpeace, Friends of the Earth, and Birdlife, support wind-energy development as a means to reduce use of fossil fuels, carbon emissions, and resulting climate change. The Nature Conservancy, for example, *supports the rapid expansion of renewable energy while protecting wildlife and natural habitats* (TNC 2022).

5.3 Wildlife—Bats

Bats are members of the order Chiroptera and comprise about one-fifth of all known mammals (~1,100 bat species out of >5,400 mammals); among mammals, only rodents have more species (Sparks et al. 2011). As with most other animals, diversity is greatest in tropical regions, where nearly half of all bat species are found. Because of their largely nocturnal habits, however, bats are poorly known and often misunderstood by many people.

Bats are the only mammals capable of true flight, for which their bodies are highly specialized. Another important ability is echolocation, which operates similar to sonar. Bats send out complex ultrasonic calls lasting 2–20 ms in duration and varying in pitch from 20 to 200 kHz (Van Ryckeghem 1998). Echos from nearby objects are detected in the bat's highly sensitive ear and interpreted for flight navigation, catching prey, and communication with other bats. Bats have surprisingly long lifespans for such small mammals; they often live 10–25 years in the wild.

Bats are the primary predators of night-flying insects, including many agricultural pest species (Sparks et al. 2011). Some bat species eat a variety of fruits, nectar, pollen, fish, and meat; only three species feed on blood. Given their high metabolic rates, bats need to eat large quantities of food and, thus, may significantly limit insect populations. They also pollinate many flowers including economically important bananas and agave. On this basis, bats are keystone species in many environments, particularly in the tropics.

While bird mortality at wind farms has gained considerable public attention, the plight of bats is less appreciated. For many people, in fact, bats fall into the same undesirable stereotype as snakes, spiders, and scorpions. The vampire legend provides further negative perception concerning bats. The unfortunate truth is that many people are afraid of bats and, thus, care little for bat conservation or the significant roles bats play in the ecosystem and for agriculture.

Large wind turbines and wind farms pose a significant danger to bats worldwide in temperate and tropical environments (Arnett et al. 2016). In North America, the hoary bat (*Lasiurus cinereus*), eastern red bat (Fig. 5.7), and silver-haired bat (*Lasionycteris noctivagans*) make up more than three-quarters of all bats killed by wind turbines (Johnson et al. 2010). According to Sparks et al. (2011, p. 18):

> *Some wind turbines kill bats in large numbers. Recent studies suggest that bats, particularly the migratory tree bats, are actually drawn to these structures where they are killed either by colliding with the blades or by barotrauma (rapid decompression that collapses the lungs).*

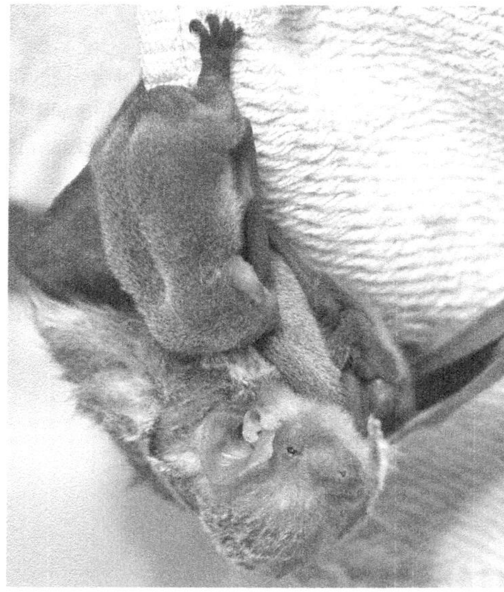

Figure 5.7 Eastern red bat (*Lasiurus borealis*) with three pups. Note the relatively large and elaborate ear. Josh Henderson/ Wikimedia Commons/CC BY-SA 2.0.

Bats in Kansas

Sparks et al. (2011) listed 16 species of bats in Kansas, which range from common to quite rare in their abundance and distribution within the state. These bats feed primarily on night-flying insects, including various flies, beetles, and moths, some of which are pests on corn, wheat, cotton, and other economically important crops. Thus, a healthy bat population is beneficial to agriculture, which is a major part of the state's economy. Bats live in and migrate through all parts of Kansas. They reside in many habitats, such as caves, old mine shafts, rocky bluffs and crevices, hollow trees and leaves, buildings, bridges, sewers, and other human-made structures.

Migratory tree bats have the greatest risk for wind-turbine mortality. The silver-haired bat resides on upper mountain slopes and in the northern plains of the United States and Canada during summer. It migrates across Kansas to spend winter in the southern United States from Arizona to the Carolinas. Eastern red bats and hoary bats are summer residents in Kansas, but likely migrate south for the winter. Most wind-turbine mortality takes place during migrations, which peak in May and September for all three bat species.

(*Continued*)

Bats in Kansas (Continued)

Bat mortality has been noted especially at the Smoky Hills Wind Farm (J. Choate, pers. com. 2008). Wooded valleys run along the base of the Blue Hills escarpment on which the wind farm is situated. Other Kansas wind farms occupy similar settings. The Red Hills region is especially important for many bat species. This region spans south-central Kansas and adjacent Oklahoma, has numerous caves, and is among the most valuable in the United States for bat biodiversity (Prendergast et al. 2010). Obermeyer et al. (2011) recommended a 15-mile (24 km) buffer around the Red Hills bat caves, and no wind farms are located in this region.

The Smoky Hills Wind Farm occupies a prominent cuesta next to lower wooded stream courses. Forest bats roost in trees along the woodland corridor during migration. Lincoln County, north-central Kansas.

The risk wind turbines pose to bats came to public attention in 2003, when hundreds of dead bats were discovered at the Mountaineer Wind Energy Center on Backbone Mountain, West Virginia (Aber et al. 2015). An estimated 1,400 to 4,000 bats were killed annually at this wind farm. Similar problems were reported soon thereafter from many other wind farms in the United States and Europe, and biologists recognized that bat mortality might be quite serious. In the European Union, in fact, all bat species are protected. The overall impact is difficult to gauge, however, because basic information is sparse or lacking about bat populations, migrations, and behaviors particularly in tropical regions.

Indiana Bat

The Indiana bat (*Myotis sodalis*) is listed as a species in danger of extinction (USFWS 2024a). These bats forage and roost in the forested areas of the eastern United States. They hibernate in caves and mines during winter, and nearly three-quarters hibernate in just four sites in Missouri, Indiana, and Illinois. The number of Indiana bats has declined to about half a million, which is only half as many since the species was first listed as endangered in the 1960s. Primary causes for reduction are human disturbance of hibernating bats, commercialization of bat caves, loss of summer habitat, pesticides, and white-nose syndrome.

The Indiana bat is among the bat species killed by wind turbines, for instance at Fowler Ridge I wind farm in northwestern Indiana. Between 2009 and 2021, a total of 34 Indiana bats were confirmed dead, usually as single individuals, at several wind farms in Pennsylvania, West Virginia, Ohio, Indiana, Illinois, Iowa, and Missouri (Pruitt and Reed 2022).

Some wind farms voluntarily shut down nighttime operations, and the companies moved to apply for Incidental Take Permits (ITP) from the U.S. Fish & Wildlife Service. An ITP *may be sought when a non-federal entity believes their otherwise lawful activities may result in take of endangered or threatened animal species* (USFWS 2024b). In particular, wind-turbine operators may deal with bat impacts by developing *Habitat Conservation Plans (HCPs) to provide for limited take of Indiana bats ... within the context of a species conservation plan* (Pruitt and Reed 2022).

Vestas V82-1.65 MW turbines in the Fowler Ridge I wind farm went online in 2009 in northwestern Indiana. During a 7-month period in that year, more than 150 bat fatalities were recorded. Most were migratory tree bats as well as a single Indiana bat (Johnson et al. 2010). Kite aerial photograph (2012).

Reducing bat mortality at wind turbines is important for bat conservation as well as public acceptance of further wind-energy development. Most bat fatalities take place in late summer and early fall during the migration and mating season in high-latitude temperate regions of North America and Europe; however, this seasonal pattern is less distinct at lower temperate latitudes and may not exist in tropical regions (Arnett et al. 2016). While migratory tree bats are most at risk, other bat species also are killed at wind turbines. What causes their attraction to wind turbines is uncertain. Nonetheless, bat behavior in Canada and the United States includes the following basic characteristics (Arnett et al. 2008; Sparks et al. 2011).

- Bats typically fly at night in calm or light-wind conditions (<6 m/s).
- Migrations typically peak in spring and late summer to early autumn.
- Bat activity is widespread during summer nights.
- Bats either hibernate or migrate southward during the winter.

On this basis, potential wind-farm bat risk could be ameliorated in three general ways.

- Wind-farm location—Determining through consultation with wildlife biologists if a proposed wind-farm location would pose a high likelihood for bat mortality. Conducting pre- and post-construction acoustic monitoring of bat activity (Arnett et al. 2016).
- Wind-turbine operation—Turning off all or selected turbines during night and low-wind conditions, when bats typically fly, particularly during peak migration periods. Setting turbines to have a cut-in wind speed that is greater than the low-wind speed favored by bats has proven effective for reducing bat mortality (Arnett et al. 2011).
- Acoustic deterrent—Broadcasting ultrasonic signals from turbines to distract or discourage bats from approaching closely. Preliminary studies have shown that such methods result in significant reduction of mortality for some bat species (e.g. Weaver et al. 2020; Good et al. 2022).

These results confirm the common-sense notion that turbines generate little electricity under low-wind operation, which happens most often during night when bats typically fly (see sect. 3.1.2). The peak migration periods correspond to reduced electricity demand between the main winter heating and summer air-conditioning seasons. Thus, electrical generation lost through these techniques would be of small concern financially and little consequence for overall electrical demand.

5.4 Visibility

Some wind farms are situated in remote locales visited and seen by few people; whereas, others are plainly visible in prominent locations and are viewed by thousands of people daily (Fig. 5.8). The aesthetic appearance of a wind farm has

Figure 5.8 Large turbines next to the heavily traveled E6/E20 motorway in southwestern Sweden are seen by thousands of people every day.

Figure 5.9 Wind turbines on the edge of Göteborg harbor with the Volvo factory complex in the background at Torslanda, southwestern Sweden. Developing wind energy in such industrial settings poses minimal environmental or aesthetic impacts.

less to do with simple visibility than how it fits into the surrounding terrain in relation to natural features and cultural objects in that windscape. Few people would object, for instance, to installation of wind turbines in a largely industrial setting (Fig. 5.9). However, those same turbines placed in a scene noted for its natural beauty could arouse considerable negative reactions from the public. Although most people now live in large cities, a rural concept of landscape identity remains deeply ingrained, which often gives rise to protectionist attitudes (Short 2002).

Among the most notorious cases is the wind-energy complex at San Gorgonio Pass next to Palm Springs, California. Interstate highway I-10 passes through the middle of the complex; average daily traffic count on I-10 in this vicinity is ~95,000 vehicles (CalTrans 2017). Wind turbines of all sizes and types once sprouted like an industrial thicket from the desert floor and hills either side of the highway (Fig. 5.10). The number of turbines has been reduced substantially in recent years. However, more than 650 turbines still operate in this vicinity (Hoen et al. 2018–2025), and their unsightly appearance remains quite obvious.

Figure 5.10 Interstate highway I-10 passes through the middle of the wind-energy complex at San Gorgonio Pass with the San Jacinto Mountains in the background, as seen in 2012. This highway is the primary link between Los Angeles and Palm Springs, California.

Given the long fetch necessary for undisturbed wind, tall turbines of wind farms may be visible from tens of km away (Twidell 2009). In relatively flat, open terrain or at sea, the curvature of the Earth is the limiting factor for visibility of tall objects from a distance, a fact long known to ship captains. For example, a turbine tower 100 m tall is potentially visible to a person standing on the horizon up to 40 km away (Young 2023). This range may vary considerably depending on atmospheric conditions and topographic settings. Wind turbines are most visible from a long distance in clear air at night when their warning lights are blinking and just after sunrise or before sunset when sun glint reflects from rotating blades. The farthest wind-farm visibility witnessed by the authors is 55 km (see sect. 7.3.3).

The concept of the viewscape, also known as intervisibility or the viewshed, pertains to visibility of objects in the landscape. More specifically it might refer to what is visible to an observer standing in a particular place, or it might relate to the visibility of a given object as seen from the surrounding terrain (Burrough 1986). Determining the viewscape begins with a digital topographic model of the terrain to which the heights of known objects are added, such as trees, buildings, and towers. In this manner the visibility of a planned wind farm could be determined and modified, if necessary, before approval and construction.

Visibility is ultimately an aesthetic issue, and truly *beauty is in the eye of the beholder* for wind farms just as for other human creations (Twidell 2009). The modern three-bladed turbine mounted on a tubular tower is at least in part a matter of public acceptable of its appearance. Still, some people consider wind turbines big, ugly machines that spoil or degrade the view. The human eye is particularly sensitive to movement even in our peripheral vision (Aber et al. 2019), and our attention is drawn to anything that moves. Active wind turbines are, thus, much more visible than are static radio towers, transmission lines, or other large stationary objects. This seems to be the fundamental objection to wind turbines rather than simply their size or height.

Saito (2004) made the distinction between "thin" and "thick" aesthetic aspects of wind farms. The thin qualities are seen in a snapshot view that takes in only the immediate windscape; whereas, the thick values include the overall environmental significance of wind energy. While most people support the environmental benefits of relatively safe, non-polluting energy harvested from the wind, the not-in-my-backyard (NIMBY) syndrome often applies when wind farms are proposed in their own neighborhoods (Good 2006). In other words, the thick values of wind farms are accepted in principle, but the local thin view is rejected.

On the other hand, well-designed and properly maintained wind farms that have closely engaged the local community during the planning process may be viewed with admiration. Developers and operators should strive to minimize the negative aesthetic impacts of wind turbines as well as the associated infrastructure of access roads, substations, and overhead power lines as much as possible (Gipe 2002).

Wind turbines are typically marked in distinctive colors to provide visual warning for low-flying aircraft and may have additional color schemes to blend in with local terrain (Fig. 5.11). In the United States, the FAA has developed

Figure 5.11 Turbines near the Danube River in eastern Austria. The blades are marked in bold red and white. Gradational green colors at the base of the tower provide a visual transition from the ground upward.

guidelines for marking and lighting wind turbines to be highly visible to aircraft both at night and in daytime (FAA Wind Turbine 2023).

- Wind turbines greater than 499 feet (152 m) above ground level require white paint and a flashing, strobe, or pulsed red obstruction light.
- Obstruction lights are placed as high as possible on the turbine nacelle and must be visible by a pilot approaching from any direction.
- Wind turbines greater than 699 feet (213 m) above ground level require an additional level of lights located midway between the top of the nacelle and ground level.
- In geographic areas that experience lengthy periods of snow cover, the mast of the turbine may be painted alternating bands of aviation orange and white.
- Wind farms less than 500 feet above ground level may qualify for reduced lighting (white paint only).

At night, wind farms have red lights on selected turbines to warn approaching aircraft, which is another visibility issue. The specific types and functioning of these lights vary from country to country (e.g. sect. 6.3.2). In the United States, the lights throughout a wind farm are supposed to blink in unison according to FAA regulations. However, meteorological towers, which are a typical component of most wind farms, may blink on a different tempo. Likewise, adjacent wind farms, radio towers, or cell-phone towers may blink out-of-sequence from each other. The net result is off-cadence blinking that creates an annoying visual distraction for night driving and for nearby rural residences.

Kansas, for example, has enacted a law that requires wind farms to have blinking red lights that activate only when radar detects an approaching airplane (Kansas SB49 2023). The technique, known as aircraft-detection lighting system (ADLS), turns red lights on when an airplane flies within three nautical miles and within 1,000 feet above the highest point of a wind farm (Fig. 5.12). Most existing and new wind farms in the state may have such systems in the future, where allowed by the FAA. Colorado and North Dakota have likewise adopted similar requirements for wind farms.

The aesthetic appreciation of man-made structures is highly subjective and may change through time. When first built, the Eiffel Tower in Paris and the Golden Gate Bridge in San Francisco were decried as ugly eyesores; both are now iconic symbols of their cities. As with all things new, early negative reactions take place. However, a paradigm shift seems to be underway for wind farms. *Initial aesthetic objections not only subsided but also changed into positive responses, embracing and celebrating the altered landscapes* (Saito 2004). Nonetheless, some objections remain for the visual appearance of wind turbines (Fig. 5.13).

Figure 5.12 GE Wind 2.82-127 turbines of the Sunflower wind project are equipped with nighttime ADLS blinking red lights. This wind farm went online in 2023 in the Flint Hills of east-central Kansas. 76 turbines have a total rated capacity of 214 MW. Hoen et al. (2018–2025).

Figure 5.13 Protest sign on US highway 50 in Marion County, east-central Kansas. The county hosts two large wind farms—Sunflower (in the background) and Diamond Vista—with a combined capacity of >500 MW. Hoen et al. (2018–2025).

5.5 Noise

Turbine noise is another issue for some people. Much research has been conducted, and field testing has shown that most turbines generate sound in the ~300–1,250 Hz range, which is the middle portion of the piano keyboard (keys 42–67). This is well below the high-frequency that poses greatest danger for human hearing. The strength or loudness of sound is measured by the decibel (dB). Zero dB is the threshold of hearing for most people, and 125 dB is the level at which sound begins to hurt. Each increase of 10 dB represents a doubling of sound level; thus, an increase of 20 dB is four times as loud (Table 5.1).

The sonic environment, or soundscape, impacts human hearing in both urban and rural settings. Many people are subjected to loud sounds on a daily basis including traffic and construction as well as agricultural and industrial noises. Some people, in fact, willingly expose themselves to extreme noise levels. Immediate hearing loss may occur from sharp, singular sounds at >140 dB, and permanent hearing loss could happen with only 5 minutes exposure to 105–110 dB, for example at loud sports or music events (Elliott 2022).

The wind generates natural background sounds that many people find pleasing, so-called white or green noise, and at increasing wind speed background sounds may mask turbine noise. Noise from wind turbines is hardly noticeable in the

Table 5.1 Common sound loudness values in decibels (dB).

Type of sound	Decibel (dB)
Whisper	30
Wind turbine	40–55*
Normal speaking	60
City traffic (inside car)	85
US train horn	96–110
Motorcycle, snowmobile, power mower	100–110
Symphonic music peak, loud rock concert	120
NASCAR, drag racing, F1 racecar, chainsaw	130–150
NFL football game peak, jet engine at 30 m	140
Firecrackers	140–150
Shot gun blast	165

* Sound limits for turbines typically required by national or local regulations.
Aber et al. (2015), USFRA (2025), Elliott (2022), WindEurope (2024), and other sources.

upwind direction, but is more audible downwind from turbines. This noise is a gentle susurration comparable to waves on the beach or wind blowing through pine trees. The typical set-back distance for wind turbines from rural residences is at least 1,000 feet (Fig. 5.14). In our experience, noise from large, modern turbines is only slightly noticeable at this distance and rarely rises above background noise levels. Modern wind turbines are designed for minimal mechanical noise and have greatly reduced aerodynamic noise (WindEurope 2024).

Most of the noise from turbines is generated by the trailing edge of the blade, particularly near the outer end of the blade which has the highest wind speed (Oerlemans 2011). Methods to reduce trailing-edge noise include optimized air-foil shape, swept blades, and serrations or brushes along the trailing edge of the blade (Fig. 5.15). On the other hand, the noise level from unmuffled gas engines

Figure 5.14 The large turbine on left is just over 1,000 feet (300 m) away from the farmstead. Vestas V120-2.2 MW turbine that stands 182 m tall with blade in upright position. Kings Point wind farm, Jasper County, southwestern Missouri (2022).

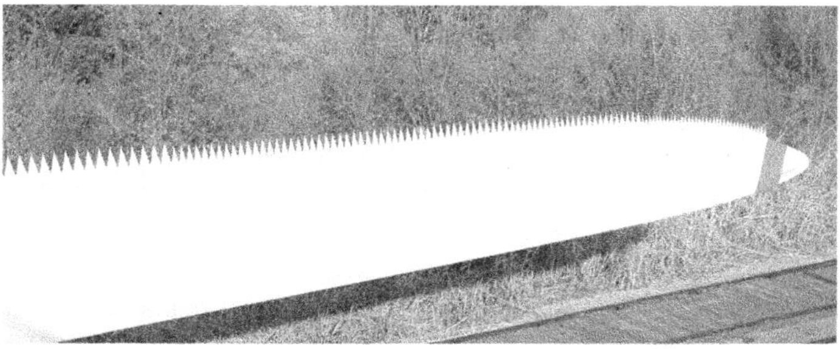

Figure 5.15 Serrated trailing edge on blade for Vestas turbine. Note the graduated size of teeth that become smaller toward the blade tip (right). Such blades are increasingly common. On a railcar in the BNSF Railway depot at Emporia, Kansas.

Figure 5.16 Unmuffled, gas-powered engine pumping groundwater from the High Plains Aquifer in westernmost Kansas. The noise resembles a race car (130–150 dB) and is painful and unbearable nearby. Thousands of such irrigation pumps are found throughout the rural Great Plains of the United States.

for pumping groundwater is far louder and much more distressful than the noise from any wind turbines (Fig. 5.16). Ultimately, visibility and noise are matters of individual perceptions and sensitivities, which vary widely from person to person.

5.6 Summary

Construction and operation of wind farms containing large turbines represent changes in land use that may have environmental and health consequences and affect the aesthetic appearance of the windscape. Clearly wind farms are not appropriate in some settings; whereas, vast areas of agricultural lands as well as shallow offshore locations are suitable for wind energy. Growth of wind energy worldwide represents an equivalent non-development of coal mines, oil and gas fields, hydroelectric projects, and nuclear energy. These energy sources often have serious environmental and cultural impacts as well as safety risks. Nonetheless, there are some negative consequences of wind energy, of which wildlife, visibility, and noise are significant issues.

The wind-energy complex at Altamont Pass, California, is among the oldest and largest collections of wind turbines in the United States and is most noted for high bird mortality. Since 2005, several steps have been taken to change the operating procedures and to retrofit the infrastructure at Altamont Pass, which have reduced bird fatalities substantially. Newer techniques of black blades and camera AI systems have shown promise for large declines in raptor fatalities in Wyoming and elsewhere.

The risk wind turbines pose to bats came to public attention early in the twenty-first century in the eastern United States and Europe. Migratory tree bats are particularly susceptible, especially during their late summer and early autumn mating and migration seasons, but many other bat species are killed by turbines in temperate and tropical environments. The Indiana bat is an endangered species that poses special concerns and has legal protection in the United States. Several techniques may be employed to reduce bat mortality by modifying turbine operation and use of acoustic deterrent methods.

Given the curve of the Earth's surface, tall turbines may be visible from tens of km away. The aesthetic appearance of a wind farm has less to do with simple visibility than how it fits into the surrounding terrain in relation to natural features and cultural objects in that windscape. Most people support the environmental benefits of relatively safe, non-polluting energy harvested from the wind, but the not-in-my-backyard (NIMBY) syndrome often applies when wind farms are proposed in their own neighborhoods. As with all things new, early negative reactions take place. However, a paradigm shift seems to be underway for public acceptance of wind farms in Europe and the United States. New ADLS may reduce annoying blinking red lights at night.

Human beings are sensitive to a wide range of sounds. Whether a sound is pleasing, annoying, or distressful, however, is often a matter of individual preference. The wind generates natural background sounds that many people find pleasing, and at increasing wind speed background sounds may mask turbine noise. This noise is a gentle susurration comparable to waves on the beach. Regulations for newer wind farms typically restrict noise levels to 40–55 dB, which resembles the human talking voice. Ultimately, visibility and noise are matters of individual perceptions and sensitivities, which vary widely from person to person.

References

Aber, J.S., Aber, S.W. and Pavri, F. 2015. Windscapes: A global perspective on wind power. Multi-Science Publ., UK, 245 p.

Aber, J.S., Marzolff, I., Ries, J.B. and Aber, S.E.W. 2019. Small-format aerial photography and UAS imagery: Principles, techniques and geoscience applications, 2nd ed. Elsevier, Amsterdam, Netherlands, 382 p.

Arnett, E.B., Baerwald, E.F., Mathews, F., Rodrigues, L., Rodríguez-Durán, A., Rydell, J., Villegas-Patraca, R. and Voigt, C.C. 2016. Impacts of wind energy development on bats: A global perspective. In: Voigt, C.C. and Kingston, T. (eds.), Bats in the Anthropocene: Conservation of bats in a changing world. Chap. 11, p. 295–323. Springer Open. https://doi.org/10.1007/978-3-319-25220-9_11

Arnett, E.B., Brown, W.K., Erickson, W.P., Fiedler, J.K., Hamilton, B.L., Henry, T.H., Aaftab, J., Johnson, G.D., Kerns, J., Koford, R.R., Nicholson, C.P., O'Connell, T.J., Piorkowski, M.D. and Tankersley, R.D. Jr. 2008. Patterns of bat fatalities at wind energy facilities in North America. *Journal of Wildlife Management* 72(1), p. 61–78. https://doi.org/10.2193/2007-221

Arnett, E.B., Huso, M.M.P., Schirmacher, M.R. and Hayes, J.P. 2011. Altering turbine speed reduces bat mortality at wind-energy facilities. *Frontiers in Ecology and the Environment* 9(4), p. 209–214. https://doi.org/10.1890/10.103

Bureau of Land Management (BLM) 2024. What makes a bird a raptor: Legal requirements for raptor possession, p. 12–13. U.S. Department of the Interior, Bureau of Land Management: Special Recreation Permits. https://www.blm.gov/sites/default/files/documents/files/Morley-Nelson-Snake-River-Birds-Of-Prey_More-About-Raptors.pdf

Burrough, P.A. 1986. Principles of geographical information systems for land resources assessment. In: Monographs on soil and resources survey. Oxford Science Publ., Vol. 12. 194 p. https://doi.org/10.1080/10106048609354060

CalTrans 2017. Traffic volumes: Route 7-10. California Department of Transportation. https://dot.ca.gov/programs/traffic-operations/census/traffic-volumes/2017/route-7-10

Center for Biological Diversity (CBD) 2004. Fact sheet on Altamont Pass bird kills. http://www.biologicaldiversity.org/swcbd/PROGRAMS/bdes/altamont/factsheet.pdf

Danish Offshore Wind 2006. Danish offshore wind: Key environmental issues. DONG Energy, Vattenfall, The Danish Energy Authority, and The Danish Forest and Nature Agency. https://tethys.pnnl.gov/sites/default/files/publications/Danish_Offshore_Wind_Key_Environmental_Issues.pdf

DHI 2018. North Sea wind power hub: Pre-screening of potential environmental impacts, Final report. DHI Business Management System. https://northseawindpowerhub.eu/files/media/document/Environmental-Pre-screening-final-report.pdf

Elliott, S. 2022. Beautifying the soundscape. *Current Affairs* 7(6), p. 45–47. https://support.currentaffairs.org/hubfs/Issues/CurrentAffairs-39.pdf?hsLang=en

FAA Wind Turbine 2023. Wind turbine marking & lighting (M&L)/aircraft detection lighting systems (ADLS). U.S. Federal Aviation Administration. https://www.faa.gov/air_traffic/flight_info/aeronav/acf/media/Presentations/23-02-ADLS-Briefing.pdf

George, G. 2023. A new study in Wyoming will test whether black paint helps birds avoid wind turbines. Audubon. https://www.audubon.org/news/new-study-wyoming-will-test-whether-black-paint-helps-birds-avoid-wind-turbines

Gipe, P. 2002. Design as if people matter: Aesthetic guidelines for a wind power future. In: M. Pasqualetti, P. Gipe and R. Righter (eds.), Wind power in view: Energy landscapes in a crowded world. Academic Press, Elsevier, p. 173–212. https://doi.org/10.1016/B978-012546334-8/50010-3

Good, J. 2006. The aesthetics of wind energy. *Human Ecology Forum* 13(1), p. 76–89. http://www.humanecologyreview.org/pastissues/her131/good.pdf

Good, R.E., Iskali, G., Lombardi, J., McDonald, T., Dubridge, K., Azeka, M. and Tredennick, A. 2022. Curtailment and acoustic deterrents reduce bat mortality at wind farms. *Journal of Wildlife Management* 86(6). https://wildlife.onlinelibrary.wiley.com/doi/abs/10.1002/jwmg.22244

Government Accountability Office (GAO) 2005. Wind power: Impacts on wildlife and government responsibilities for regulating development and protecting wildlife. GAO-05-906. http://www.gao.gov/new.items/d05906.pdf

Hoen, B.D., Diffendorfer, J.E., Rand, J.T., Kramer, L.A., Garrity, C.P. and Hunt, H.E. 2018. United States Wind Turbine Database V8.0 (February 25, 2025). U.S. Geological Survey, American Clean Power Association, and Lawrence Berkeley National Laboratory data release. https://energy.usgs.gov/uswtdb/ or https://doi.org/10.5066/F7TX3DN0

Johnson, G., Ritzert, M., Nomani, S. and Bay, K. 2010. Final report bird and bat fatality studies Fowler Ridge I wind-energy facility Benton County, Indiana: April 6 – October 30, 2009. Western Ecosystems Technology Inc. https://tethys.pnnl.gov/publications/final-report-bird-bat-fatality-studies-fowler-ridge-i-wind-energy-facility-benton

Kansas SB49 2023. SB 49: Requiring installation of light-mitigating technology systems on new and existing wind energy conversion systems subject to certain conditions. Kansas Legislature. https://www.kslegislature.gov/li/b2023_24/measures/sb49/

Kusnetz, N. 2021. Canada's tar sands: Destruction so vast and deep it challenges the existence of land and people. Inside Climate News. https://insideclimatenews.org/news/21112021/tar-sands-canada-oil/

Lowitz, M. 2019. Encyclopedia of Earth. Altamont Pass, California. http://editors.eol.org/eoearth/wiki/Altamont_Pass,_California

Lynes, M. 2023. Audubon settlement improves protections for golden eagles in California's Altamont Pass. Audubon. https://www.audubon.org/news/audubon-settlement-improves-protections-golden-eagles-californias-altamont-pass

NDOT 2013. New painted contrast markings increase visibility, save money. Roadrunner, Nebraska Department of Transportation, August/September, 2013. https://dot.nebraska.gov/media/r4jpdbo0/new-contrast-markings.pdf

NSWPH 2022. Hubs and spokes – Viable beyond theory. North Sea Wind Power Hub Programme. Concept Paper 2022. https://northseawindpowerhub.eu/knowledge/hubs-and-spokes-viable-beyond-theory

Obermeyer, B., Manes, R., Kiesecker, J., Fargione, J. and Sochi, K. 2011. Development by design: Mitigating wind development's impacts on wildlife in Kansas. *PLoS One* 6(10), e26698. https://doi.org/10.1371/journal.pone.0026698

Oceanographic Staff 2022. New research shows 98% decrease in bottom-towed fishing in Dogger Bank MPA since ban. Oceanographic Magazine. https://oceanographicmagazine.com/news/new-research-shows-98-decrease-in-bottom-towed-fishing-in-dogger-bank-mpa-since-ban/

Oerlemans, S. 2011. Chapter 2. Primary noise sources. In: D. Bowdler and G. Leventhall (eds.), Wind turbine noise. Mulit-Science Publ. Co., Brentwood, Essex, United Kingdom, p. 13–45.

Phillips, E., Cotterill, C., Johnson, K., Crombie, K., James, L., Carr, S. and Ruiter, A. 2017. Large-scale glacitectonic deformation in response to active ice sheet retreat across Dogger Bank (southern central North Sea) during the Last Glacial Maximum. *Quaternary Science Reviews* 179, p. 24–47. https://doi.org/10.1016/j.quascirev.2017.11.001

Prendergast, J.A., Jensen, W.E. and Roth, S.D. 2010. Trends in abundance of hibernation bats in a karst region of the southern Great Plains. *The Southwestern Naturalist* 55(3), p. 331–339. https://www.jstor.org/stable/40801031

Pruitt, L. and Reed, M. 2022. Indiana bat fatalities at wind energy facilities. U.S. Fish & Wildlife Service, Indiana Field Office. https://www.fws.gov/sites/default/files/documents/Indiana Bat Fatalities at Wind Energy Facilities_081622.pdf

Saito, Y. 2004. Machines in the ocean: The aesthetics of wind farms. *Contemporary aesthetics* 2(6). https://digitalcommons.risd.edu/liberalarts_contempaesthetics/vol2/iss1/6/

Short, L. 2002. Wind power and English landscape identity. In: M. Pasqualetti, P. Gipe and R. Righter (eds.), Wind power in view: Energy landscapes in a crowded world. Academic Press, Elsevier, p. 43–58.

Sparks, D.W., Schmidt, C.J. and Choate, J.R. 2011. Bats of Kansas. Indiana State University, Center for North American Bat Research and Conservation, Pub. No. 5, 60 p. https://scholars.fhsu.edu/cgi/viewcontent.cgi?article=1007&context=sternberg_facpubs

Thelander, C.G., Smallwood, K.S. and Rugge, L. 2003. Bird risk behaviors and fatalities at the Altamont Pass wind resource area. National Renewable Energy Laboratory, NREL/SR-500-33829. http://www.nrel.gov/docs/fy04osti/33829.pdf

Tinker, S.W. 2014. Energy 360: Moving energy – No easy feat, but a vital conversation. *Earth* 59(1), p. 8–9. https://www.earthmagazine.org/article/comment-energy-360-let-science-speak-not-agendas/

TNC 2022. Site renewables right: Accelerating a clean and green renewable energy buildout in the central United States. The Nature Conservancy. https://www.nature.org/content/dam/tnc/nature/en/documents/SRR_Methods_20220202_LR.pdf

Twidell, J. 2009. Fundamentals of wind turbines. In: J. Twidell and G. Gaudiosi (eds.), Offshore wind power. Mulit-Science Publ. Co., Brentwood, Essex, United Kingdom, p. 15–26.

USFRA 2025. Train horns and quiet zones. U.S. Department of Transportation, Federal Railroad Administration. https://railroads.dot.gov/railroad-safety/divisions/highway-rail-crossing-and-trespasser-programs/train-horn-rulequiet-zones

USFWS 2017. Threats to birds. U.S. Fish & Wildlife Service. https://www.fws.gov/library/collections/threats-birds

USFWS 2024a. Indiana bat. U.S. Fish & Wildlife Service. https://www.fws.gov/species/indiana-bat-myotis-sodalis

USFWS 2024b. Permits for native endangered and threatened species. U.S. Fish & Wildlife Service. https://www.fws.gov/library/collections/permits-native-endangered-and-threatened-species

Van Ryckeghem, A. 1998. How do bats echolocate and how are they adapted to this activity? Scientific American, Ask the Experts. http://www.scientificamerican.com/article/how-do-bats-echolocate-an/

Weaver, S.P., Hein, C.D., Simpson, T.R., Evans, J.W. and Castro-Arellano, I. 2020. Ultrasonic acoustic deterrents significantly reduce bat fatalities at wind turbines. *Global Ecology and Conservation* 24. https://doi.org/10.1016/j.gecco.2020.e01099

Wells, J. 2021. How Duke Energy is using technology to save eagles at a Wyoming wind site. Illumination. Duke Energy. https://illumination.duke-energy.com/articles/how-duke-energy-is-using-technology-to-save-eagles-at-a-wyoming-wind-site

Wiegand, J. 2013. Wind turbines kill up to 39 million birds a year! CFACT.org. https://www.pichimahuida.info/energy_files/Wind turbines kill up to 39 million birds a year! - CFACT.pdf

WindEurope 2024. Wind energy and the environment. WindEurope. https://windeurope.org/about-wind/wind-energy-and-the-environment/

Young, A.T. 2023. Distance to the horizon. http://mintaka.sdsu.edu/GF/explain/atmos_refr/horizon.html

6

Europe and Asia

6.1 Eurasian Overview

Europe and Asia together form the world's largest continental landmass, namely Eurasia. Geographic features include the highest (Mt. Everest) and lowest (Dead Sea) places on land, an extreme environmental range from tropical forest to permafrozen tundra, great mineral wealth, and dense human population to sparsely settled regions. Eurasia contains a majority of the world's people as well as the greatest overall development of wind energy. China, Germany, India, and Spain represent four of the five largest wind-energy countries with a combined capacity of 586 GW, more than half the global total (see Table 1.3). Likewise wind turbine manufacturers in China, Denmark, Spain, and Germany dominate global production (see Table 1.4).

The degree of wind-energy generation spans a broad range country by country from intensive development, for instance Denmark, Spain, and Portugal, to negligible, as for Indonesia, Kuwait, Slovakia, and Slovenia (Wind Power Countries 2025). The world's largest country, Russia, surprisingly has insignificant wind energy, presumably because of its huge petroleum reserves. Across this vast and diverse megacontinent, case studies are elaborated for Denmark, Faroe Islands, Poland, and India representing large contrasts in geographic conditions, historical development, natural resources, and modern cultures and economies.

Wind Energy Landscape: Principles, Techniques, History, First Edition. James S. Aber, Susan E. W. Aber, and Firooza Pavri.

Published 2026 by John Wiley & Sons, Inc.

6.2 Denmark

6.2.1 Introduction

Denmark is a world leader in so-called green energy. Denmark in 2022 already reached 60% of electricity production from renewable sources—wind, solar, and hydro power (Berggreen 2023). The country has the goal to achieve self-reliance on renewable energy by 2050 (DEA 2015). Denmark's wind-energy capacity achieved 7.1 GW in 2023 (Wind Power Denmark 2025), and given a population of nearly six million people, this gives the country the highest per-capita wind-power capacity in the world (Ember 2024).

Denmark has more than 1,800 listed wind farms, but most of these have just one to <10 turbines. This reflects the typical small-scale development that was undertaken by individuals and local cooperatives during the late twentieth century. The emphasis during the twenty-first century shifted to construction of larger wind farms both offshore and on land. Furthermore, many of the listed older wind farms have been dismantled, which follows government policy to replace many, smaller, low-capacity turbines with fewer, larger, higher-capacity turbines. Onshore wind farms from northwestern Denmark are described in detail below (Fig. 6.1).

6.2.2 Udbyneder

Udbyneder wind farm is located in a coastal wetland facing the Kattegat (sea) north of Aarhus. The wind-farm vicinity is underlain by upper Cretaceous chalk bedrock similar to chalk found in many other parts of Denmark. Surficial sediment

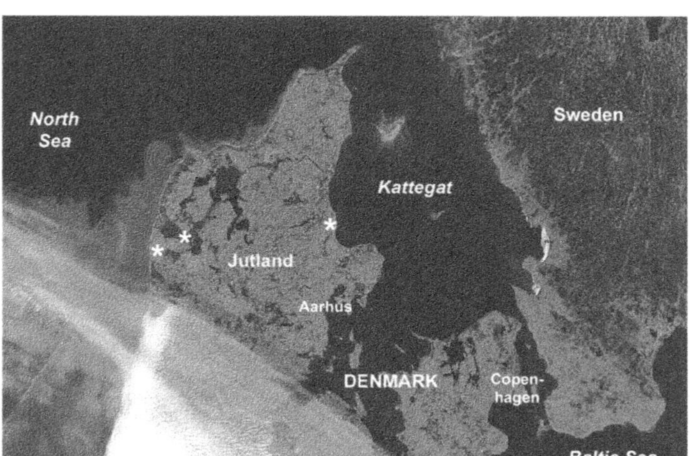

Figure 6.1 Denmark and surrounding land and sea areas. Selected onshore wind farms (*) from west to east: Ramme, Oddesund, and Udbyneder. NASA/Public Domain

Oddesund Nord, Denmark

Oddesund is a narrow finger of land that separates inland waters of Nissum Bredning from the Limfjord in northwestern Denmark. The northern peninsula at Oddesund Nord is underlain by sand and gravel that were deposited as beach ridges to form a narrow spit. This area is now a nature preserve, part of the UNESCO Global Geopark Vestjylland (Geopark 2025). But this place once appeared quite different. In 1986, Vestas and Bonus built the first large wind farm in Europe. Twenty 55 kW turbines were erected here, and 25 more 95 kW turbines were placed on the southern side of the Oddesund.

These turbines were erected in a coastal wetland and were located in clear view beside highway 11, a major touristic route leading into the summer vacationland of northwestern Jutland. The turbines stood for more than two decades, but in 2008 they were considered redundant and were removed. From the geopark observation tower, the former wind-farm site appears to be a pristine marsh and wet-meadow environment. An aerial view, however, reveals traces for some of the former turbine sites. Nonetheless, this vicinity

Former wind farm at Oddesund. The steel-truss towers and transmission lines gave an industrial appearance. Undated picture obtained and adapted from a public display at Oddesund Nord tower; original color photograph by Ib Bj. Sørensen.

(*Continued*)

Oddesund Nord, Denmark (Continued)

Overview of Oddesund Nord looking westward. Nissum Bredning in the left background, and highway 11 on the right. Sites of two former wind turbines are marked (*). Kite aerial photograph (2023).

has been restored and is quite popular for the public to wander along the beach and hike through the wet-meadow habitat. Oddesund Nord is a good example of how obsolete wind farms may be reclaimed to near-natural conditions.

at Udbyneder consists of unconsolidated marine deposits of sand and shells that were laid down in the middle Holocene a few 1000s years ago and cover the area up to 5 m above present sea level (Pedersen et al. 1989).

The area of the wind farm is just a few meters above sea level and completely open to wind blowing inland from the Kattegat (Fig. 6.2). The fetch (length) eastward across the Kattegat to Sweden is ~150 km. Given a consistent easterly wind, such a long fetch across open water leads to faster speeds and more power in the wind. From other directions, wind passes over low-lying land that is devoted mostly to pasture and crop agriculture. On this basis, the land-based Udbyneder wind farm has a low surface roughness (α) in the range 0.1 to 0.2 (see Table 3.2) and, thus, closely approaches offshore conditions.

The Udbyneder wind farm was constructed in multiple phases beginning in 2002 with Neg Micon NM72/2000, then Siemens SWT-2.3-93, and finally Vestas V126/3450 models (Wind Power Udbyneder 2025). However, many of the older and smaller Neg Micon turbines were dismantled in 2020. The current total capacity of the wind farm is 62.3 MW from 18 operational turbines. All electric cables are buried and feed into a local substation.

The low-lying vicinity is drained by a network of ditches and is used primarily for various agricultural crops and hay production. The turbines are arrayed in a

Figure 6.2 Udbyneder wind farm looking east toward the Kattegat (sea) on the horizon. Siemens turbines in the foreground and to right; Neg Micon turbines in the background. Note the grid arrangement of turbines along field boundaries and drainage ditch. Kite aerial photograph (2013).

grid pattern along field boundaries and serviced by gravel paths that result in minimal loss or disruption of existing cropland use. The wind farm is located far from major highways, tourist attractions, or sensitive environmental sites.

6.2.3 Ramme

Ramme wind farm is located in an agricultural upland approximately 5 km inland from the North Sea coast near Bovbjerg cliff. The cliff stands 44 m high and is composed entirely of unconsolidated and deformed glacial sediment. The cliff marks the terminal moraine built by the last major ice sheet to reach Denmark about 22,000 years ago (Houmark-Nielsen 2007). The moraine extends inland as a prominent ridge, an ideal location for wind turbines to catch westerly winds.

Ramme is a prime example of upgrading for Danish wind farms over the years. The wind farm began with Vestas V27 and Bonus models. These relatively small turbines were replaced later with Vestas V42 and V80 turbines installed in 1996, 2000, and 2009 (Fig. 6.3). Many of these were replaced in turn with still larger Vestas models. As of 2024, Ramme wind farm had 12 Vestas turbines, the largest of which has a capacity of 3.3 MW. Total combined capacity of the wind farm is 31.8 MW (Wind Power Ramme 2025). In addition to turbines, the wind-farm vicinity recently added a large array of solar panels (Fig. 6.4).

The Ramme wind–solar complex is located within clear view of a well-known archaeological site, Ramme Dige. It includes a cluster of burial mounds dating from the Stone Age and Bronze Age, as well as an Iron Age wall-ditch structure that extends more than 2½ km long (see Fig. 2.10). Both Ramme Dige and Bovbjerg cliff are highly popular tourist stops in northwestern Jutland, but the presence of the wind–solar complex seems to have minimal detrimental effect on tourism or surrounding agricultural land use.

Figure 6.3 Ramme wind farm looking toward the northeast in the rural landscape of western Jutland, Denmark, as it appeared in 2013. Trace of an Iron Age wall-ditch is marked (*). Compare with Fig. 2.10; kite aerial photograph.

Figure 6.4 Array of solar panels (dark gray) partly surrounds turbines at the northern end of the Ramme wind farm. Kite aerial photograph (2023).

6.2.4 Danish Energy Status

Denmark's near-term target is to reach 40 GW of renewable capacity by 2030. This ambitious goal is to provide for more climate-friendly electricity and heating and to secure energy independence from unreliable foreign sources, namely Russia. In addition to wind power, Denmark is moving rapidly to develop solar energy, and solar farms have sprung up in many places (Fig. 6.5). During the decade from 2013 to 2022, annual solar-energy production increased from near zero to 2,188 MWh (Berggreen 2023). This is still small compared with wind-power production of >19,000 MWh in 2022, but solar farms may be situated within existing wind farms or in many sites that would be unfavorable for wind turbines.

Another major part of this effort began in 2020 to construct two so-called energy islands (Energinet 2020). These energy islands represent a new epoch for large-scale, offshore wind power (Energinet 2023). One energy island is in the North

Figure 6.5 Solar-energy photovoltaic (PV) array for generating electricity. Island of Møn, southeastern Denmark. Kite aerial photograph (2023).

Sea ~80–100 km west of Thorsminde on the west coast of Jutland. Initial capacity is planned for 3 GW, which could expand to 10 GW in the 2030s. The artificial island is essentially an electrical collection station that receives AC current from surrounding offshore wind turbines. The AC current is transformed into DC current, which will be sent via undersea cables to Denmark and other countries—Germany, Netherlands, Norway, and United Kingdom.

Bornholm is the other energy island, the easternmost part of Denmark situated in the Baltic Sea between Sweden and Poland. It will have a 2-GW capacity, and it will be the collection point for AC electricity from offshore turbines in the surrounding sea. Buried land cables and a transformer station (to DC current) will be located on the southern edge of the island. Bornholm will include undersea cable connections to eastern Denmark and Germany as well as to Sweden and Poland.

Known as the *Pearl of the Baltic Sea*, Bornholm has a lively tourism industry based on its history, nature, and rustic conditions. Development of energy-island infrastructure would avoid those areas known for tourism, for example, the world-famous, white-sand beach and dunes of Dueodde (Fig. 6.6).

Figure 6.6 Overview of the Baltic Sea, beach, dunes, and forest looking toward the east at Dueodde, southeastern tip of Bornholm. Kite aerial photograph (2023).

6.3 Faroe Islands

6.3.1 Introduction

The Faroe Islands are a self-governing part of the Kingdom of Denmark. The Faroes comprise a cluster of 19 small islands in the North Atlantic Ocean at roughly 7° W longitude and 62° N latitude about halfway between Scotland and Iceland. The climate is surprisingly mild for this high latitude owing to the influence of warm water in the Gulf Stream. Nonetheless, the Faroes have a reputation for abundant cloud cover, frequent rain, and stormy weather.

The Faroe Islands are remnants of a volcanic plateau that once resembled Iceland some 60 million years ago. However, erosion and subsidence have reduced the islands greatly to just 1,400 km^2 land area (Rasmussen 1982). The landscape is quite rugged overall with steep slopes and little arable land (Fig. 6.7). Other than thin coal seams that were mined in the past, all fossil-fuel energy resources must be imported.

Human population has grown rapidly in the twenty-first century and now exceeds 54,000 people, of which more than 20,000 live in the capital region in and around Tórshavn. The islands also are home to about 80,000 sheep. A thriving fishing industry and tourism, especially for bird watching, contribute to the economy of the islands.

Rapid population growth in recent years means more electricity is necessary to support an increasing standard of living. The Faroe Islands, like many relatively small and isolated islands, are well positioned to take advantage of available renewable energy resources (Barney et al. 2022). The Faroes already produce more than half of the nation's electricity from renewable sources including hydropower, wind turbines, and tidal power (Faroe Energy 2025). However, fossil fuel is still important for producing electricity as well as for home heating and powering the vast fishing fleet.

Figure 6.7 Typical landscape in the Faroe Islands. Village of Viðareiði on the island of Viðoy, northernmost Faroe Islands. Villingadalsfjall mountain peak is 841 m above sea level.

The so-called "mainland" consists of the central and northeastern islands. The mainland represents about 90% of electricity use overall. Not included are several small islands and the southern island of Suðuroy. These other islands have their own electricity sources. Most electricity is produced currently by oil, hydropower, and wind energy. An online app displays real-time energy usage overall and for individual wind farms, hydro plants, etc. (SEV 2025).

The Faroe Islands have set the goal to achieve 100% renewable electricity by 2030, and half of all households should be heated by renewable energy also by 2030. SEV, the Faroese electric utility, received a Nordic Council Nature and Environment Prize in 2015 for setting and working toward these goals (Faroe Energy 2025). Wind energy undoubtedly will play a prominent role.

6.3.2 Wind Energy

The Faroe Islands have a long history with wind energy, since Christian Riisager moved to Miðvágur in 1981 and continued building wind turbines there (Thorndahl 2011). The islands currently have seven onshore wind farms (Wind Power 2025). The largest of these are the Húsahagi, Tórshavn, and Flatnahagi wind farms with 25 turbines together on the mountain plateau just west of the capital of Tórshavn on the island of Streymoy. The prevailing winds there are from the southwest and west, which have the greatest frequencies as well as highest average speeds at 10–20 m/s (Cappelen and Drost Jensen 2021). Furthermore, mean wind speed is greatest during the autumn and winter months (November–March), when more electricity is needed for heating and lighting.

Three generations of wind turbines are present including Enercon and Vestas models, the largest of which are Vestas V117/4000–4200 turbines that are designed for extreme wind conditions (Fig. 6.8). Smaller, older turbines spin faster than

Figure 6.8 Portion of the Húsahagi and Tórshavn wind farms on the upland plateau more than 300 m above sea level. Enercon turbines in the foreground and Vestas turbines in the background.

Figure 6.9 Electric substation in the Húsahagi wind-energy park beside highway 50. Reflective glass panels obscure the electrical apparatus behind.

larger, newer turbines. At night, each turbine has two non-blinking red lights on top of the nacelle.

The Húsahagi wind-energy park straddles highway 50, and the turbines are clearly visible along the highway as well as from many other places in the Tórshavn vicinity. To minimize the visual impact, all electric cables are buried and a low-profile substation is shielded with glass panels that reflect the sky and clouds (Fig. 6.9). In 2018, the local utility announced plans to add 150 MW of wind power by 2030, of which 48 MW were implemented by 2021. But since then little further progress has been made due to a combination of complaints, disputes between stakeholders, and slow bureaucracy (Barney et al. 2022).

6.3.3 Green-Energy Status

In order to meet the daunting goal of 100% renewable energy for electricity by 2030, several resources and technologies will be necessary including wind, hydro, tidal, geothermal, solar, and biogas. Hydropower plants, for example at Vestmanna, are pump-storage systems, in which excess electricity is used to pump water up into reservoirs. This is a kind of "hydro battery" for storing potential energy to generate electricity when needed. Hydropower has great potential to provide reliable electricity, but strong public resistance exists for construction or enlargement of dams. The Faroes also have an ideal situation for harvesting tidal energy.

Few onshore locations are available for installing new wind farms without disturbing nearby inhabitants or potentially impacting tourism. Already resistance has formed against further expansion of onshore wind energy. Offshore wind energy has great potential, but is as yet unrealized. Geothermal energy is utilized already for home heating systems both for new homes and retrofitting older homes. Solar energy also may play a minor role, although the islands are well known for cloud

Tidal Power, Faroe Islands

Tidal energy is reliable and would be available more or less continuously, because timing of tides varies from channel to channel between the Faroe Islands and from north to south. The Dragon 4 undersea tidal-power facility was completed and grid-connected in Vestmannasund between Streymoy and Vágar islands in 2022 (Minesto 2025). Electricity is generated by submerged vehicles called "kites" that are tethered to the seafloor and move in a figure-eight pattern at 6–10 m/s. Four additional tidal-kite arrays are planned each with 20 to 40 MW capacity. Tidal-power capacity would total 120 MW and would generate an estimated 350 GWh per year.

Side view of the *Íðunn* tidal-power kite resting in a carriage on the dock at Vestmanna. Each kite is 5 m wide and weighs 1.7 tons.

cover, and high latitude would limit available sunlight for half the year. When economic and technical viability are considered as well as local environmental and social impacts, a clear preference has emerged for future offshore development of renewable wind and tidal energy for the Faroe Islands (Barney et al. 2022).

6.4 Poland

6.4.1 Introduction

Poland experienced rapid growth in wind energy early in the twenty-first century, increasing from a trivial 5 MW capacity in 2000 to more than 9 GW of onshore capacity at the beginning of 2024, and this capacity was predicted to more than

Figure 6.10 Reconstructed windmill near Czarnków, west-central Poland. Typical post mill with four blades formed by wooden frames over which cloth was stretched to form sails. Mill body rested on a stout post and could be rotated around the central post to face in any direction. Like nearly all traditional European windmills, the blades rotated in a counterclockwise direction as viewed from the upwind (front) side. Christensen and Thorndahl (2012). This mill was renovated with support from the European Union Agricultural Fund for Rural Development (2007–2013).

double by 2030 (Strzelecki and Evans 2024). The first offshore wind farm in the Baltic Sea was anticipated to begin construction in 2025.

In total, Poland had 355 wind farms in service, under construction, or planned with a total generating capacity of more than 18 GW (Wind Power Poland 2025). The greatest numbers of onshore wind farms are located along the northern Baltic coastal region and in west-central Poland. The latter has a long history of windmills as shown by the reconstructed post mill near Czarnków (Fig. 6.10). The following example comes from the same region in west-central Poland.

6.4.2 Radwanki-Margonin

The Radwanki-Margonin wind farm is located in the province of Greater Poland (Wielkopolska), which is considered the core of the earliest Polish nation. The landscape is relatively flat and devoted mainly to agriculture with scattered forests, lakes, and river valleys. Named for the village of Radwanki and nearby city of Margonin, the wind farm is situated amid agricultural fields that are bounded by forested tracts.

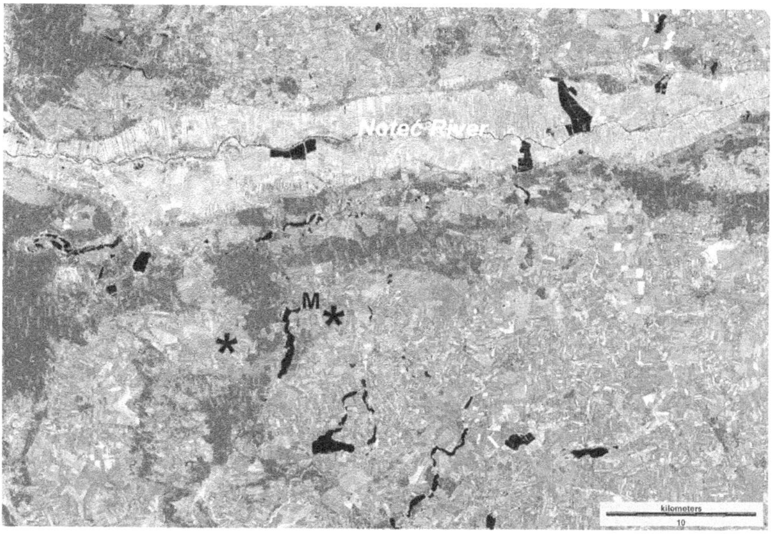

Figure 6.11 Satellite image of the Margonin (M) vicinity, west-central Poland. Asterisks indicate locations of the wind farm—villages of Radwanki (west) and Kowalewo (east). Landsat TM 345 false-color composite: forest = dark green or dark gray; agricultural land and urban = yellow-green, pink and purple or light gray; water bodies = black. Acquisition date: July 12, 2010. Dataset provided from the EROS Data Center, U.S. Geological Survey.

The most prominent geographic feature of this region is the Noteć River valley that trends east-west a short distance north of the wind-farm vicinity (Fig. 6.11). This valley was a product of glaciation during the Pleistocene Epoch (Ice Age). Like Denmark, Poland experienced multiple episodes of ice-sheet glaciation, the last of which advanced over the northern portion of the country about 20,000 years ago (Marks 2004). As the glacier margin retreated northward, the Noteć River valley was uncovered and became the primary ice-marginal drainage route for glacial meltwater leading westward beginning around 15,000 years ago (Ehlers and Gibbard 1996).

The wind farm occupies high ground on the drainage divide southward from the Noteć River valley, a site well known for favorable wind conditions as demonstrated by several older turbines in the vicinity (Fig. 6.12). The substratum consists of glacial deposits laid down during successive ice-sheet advances. Agricultural cropland is based primarily on till, namely boulder-clay glacial sediment; whereas, forested tracts occupy areas of sandy soil from meltwater deposition. Farms in Poland are of two general types. Large cooperative enterprises are derived from collective farms of the Communist era; whereas, small fields that remained in private ownership appear as a patchwork of narrow stripes. Both

Figure 6.12 Older turbines present within the Radwanki-Margonin wind farm. Smaller Nordex (N26 or N27) turbines of 1990s vintage on truss towers with a newer Gamesa G90/2000 turbine. Near the village of Kowalewo, as seen in 2013.

Figure 6.13 Typical agricultural scene in the gently undulating till landscape near the Radwanki-Margonin wind farm. Large fields of a cooperative farm on left; narrow stripes on right show fields in private ownership. The forest in the background borders the southern margin of the Noteć River valley. View northward; kite aerial photograph (2013).

patterns of agricultural land use are present within the area of the wind farm (Fig. 6.13).

The wind farm was commissioned in 2009 with 60 Gamesa G90/2000 turbines; rotor diameter 90 m, nominal capacity 2.0 MW, total capacity 120 MW (Wind Power Margonin 2025). Turbines are scattered in agricultural fields surrounding the villages of Radwanki and Kowalewo. Beyond the villages, few trees or human structures are present in proximity to the turbines; adjacent forest is sufficiently far away so as not to impede near-surface wind (Fig. 6.14). Within the area of the wind farm, all electric cables are buried, and signage warns of potential dangers.

6.4.3 Polish Economic and Energy Issues

Poland's overall economy has been liberalized since the breakup of the Soviet empire in 1990, and Poland was the only country within the European Union (EU) to avoid recession during the 2008–2009 economic downturn. Its high-growth economy did suffer a minor recession, however, during the COVID-19

Figure 6.14 Gamesa G90/2000 turbines in the Radwanki-Margonin wind farm, Poland. Hub height is 80 m. Note the open aspect of the site west of Radwanki; forest margin in the background is approximately 1.5 km away from the turbines. View southward; kite aerial photograph (2013).

Figure 6.15 Bituminous coal stockpile in preparation for winter heating behind a building in Warsaw, Poland. Coal was burned in the past in building furnaces without any pollution control.

pandemic (CIA Poland 2025). Nearly one-half of the country is used for agriculture; food and live animals are exported primarily to Germany and other European countries. The Radwanki-Margonin wind farm demonstrates development and operation compatible with maintaining such agricultural production.

Poland is heavily dependent on its large domestic coal reserves for generating electricity and heating. Bituminous (hard) coal is mined in the southern part of Poland, and brown coal (lignite) is produced from large strip mines in the central portion of the country. Poland currently produces 70% of its electricity from coal (Strzelecki et al. 2023). Burning coal results in significant air pollution, particularly during the winter heating season (Fig. 6.15). On the other hand, Poland imports most of its oil and natural gas, primarily from Russia.

Wind energy grew rapidly in Poland during the early twenty-first century, as noted above. However, development of wind farms took place with few restrictions.

Figure 6.16 Looking to the southeast over the village of Radwanki. The center turbine in the background is only 400 m from the nearest house in the village. Kite aerial photograph (2013).

This largely unregulated expansion of wind energy led to considerable negative public reaction, as turbines appeared in proximity to villages and inhabited places (Fig. 6.16). The government of the conservative Law and Justice (PiS) party reacted in 2016 with the so-called 10H rule for new wind projects (Fedorska 2023). This rule required that a wind turbine must be placed a distance of at least 10 turbine heights from the nearest house or nature preserve. For a turbine with total height of 150 m, for example, this meant a setback of 1.5 km. Given the relatively dense settlement of the countryside, this effectively ended construction of new wind farms.

Russia's invasion of Ukraine in 2022 created a Polish energy crisis, which was exacerbated by a decline in European hydro- and nuclear-power generation. This led to a proposal to reduce the onshore turbine setback limit to 500 m; however, parliament could not agree. A compromise 700-m-setback was accepted finally and signed by the President in 2023 (ITA 2023). This would open an estimated 4–5% of the countryside for additional wind-power development.

Late in 2023, the incoming pro-European government submitted a draft law to parliament to make rules easier to build wind farms. For example, the turbine setback distance would be reduced to only 300 m for residential areas depending on noise levels (Strzelecki et al. 2023). The new government aimed to speed up the transition from coal and to reach 70% renewable electricity production by 2030. However, these plans soon had to be abandoned due to a strong backlash that rules would be too favorable for the wind industry (Barteczko and Krasuski 2023). Thus, the 700-m-setback distance remains in force for any new onshore wind turbines (TPA 2025).

In the meantime, Poland plans to spend $27 billion to develop offshore wind-energy resources and necessary logistical and infrastructure by 2040 (ITA 2021). The Port of Gdynia would serve as the land hub for offshore electricity generation. A dozen offshore wind farms with more than 8.5 GW capacity were planned for the Baltic Sea as of early 2024 (Wind Power Poland 2025), as well as an interconnection to the Danish energy-island Bornholm (see sect. 6.2.4).

Four Russian nuclear plants that were under construction in the late 1980s were canceled in 1990 when Poland became independent. Thus, Poland had no nuclear-power plants in operation. To mitigate its energy issues and to continue economic development Poland is turning once again to nuclear energy. In 2021, Poland committed to building up to six nuclear-power plants, limiting coal for generating electricity, using natural gas as a "bridge fuel," and increasing renewable resources to about one-third of electricity generation by 2030 (Nuclear Poland 2025).

Governmental agencies and companies from Germany, South Korea, and the United States are involved in partnerships with the state-owned company, Polish Nuclear Power Plants (Polskie Elektrownie Jądrowe, PEJ). In 2023, Poland's Ministry of Climate and Environment gave the go-ahead for PEJ to construct the first nuclear-power plant at the Lubiatowo-Kopalino site, northwest of Gdynia near the Baltic Sea (Nuclear Poland 2025). Construction would begin presumably in 2026 with completion set for 2033. Poland's nuclear-energy program is proceeding with strong public support; surveys have found up to three-quarters of Poles favor the use of nuclear power.

Overall, Poland lacks a coherent national vision for transforming from fossil fuels to clean and green energy resources (Kardaś 2023). This slow pace of reducing carbon emissions has led to loss of substantial amounts of EU funding. Nuclear power certainly would reduce the need for fossil fuels in the long term. Given the political stalemate for onshore wind turbines, offshore wind farms represent the best avenue for Poland quickly to expand green energy, to decrease reliance on coal, and to reduce carbon emissions. Nonetheless, a full transition from coal and other fossil fuels to wind and nuclear energy would take at least two or more decades.

6.5 India's Wind Geography

6.5.1 Indian Overview

India's extensive wind-energy potential varies significantly across different geographical regions due to diverse climatic and topographical conditions (Fig. 6.17). Monsoonal wind patterns play major seasonal roles and influence the geography of wind across the sub-continent. India's National Wind Resource Assessment

Figure 6.17 Typical wind turbine in India. This appears to be an Enercon turbine, perhaps the E-70 model (Wind-Turbine-Models 2025). Syedsalim1109/Wikimedia Commons/ CC BY-SA 4.0.

program has installed hundreds of monitoring stations nationwide, to help ensure an accurate and detailed mapping of India's sustained wind resources at various heights above ground level (NIWE 2023).

Wind-energy generation for India has increased from over 27,000 GWh in 2014 to nearly 69,500 GWh in 2023 (IRENA 2024). Globally India ranks fourth with an installed wind-energy capacity of 44.7 GW in 2023 (see Table 1.3). Although wind power accounted for a modest 4% of the electricity generated nationally in 2023, it is projected to rise significantly, reaching about 10% by 2032, contingent upon meeting policy and investment targets (Shah 2024).

India's consistent year-round wind-rich regions predominantly lie in the country's western and southern portions, with the highest capacities concentrated in the states of Andhra Pradesh, Gujarat, Karnataka, Maharashtra, Rajasthan, and Tamil Nadu. Among these, Tamil Nadu and Gujarat have led the development of wind-energy generation over the past decade and more, contributing nearly 50% of the country's wind-energy installed capacity (Fig. 6.18; MNRE 2025b). India also exhibits considerable offshore wind-energy potential, notably along the coastlines of Gujarat and Tamil Nadu, where initial assessments indicate

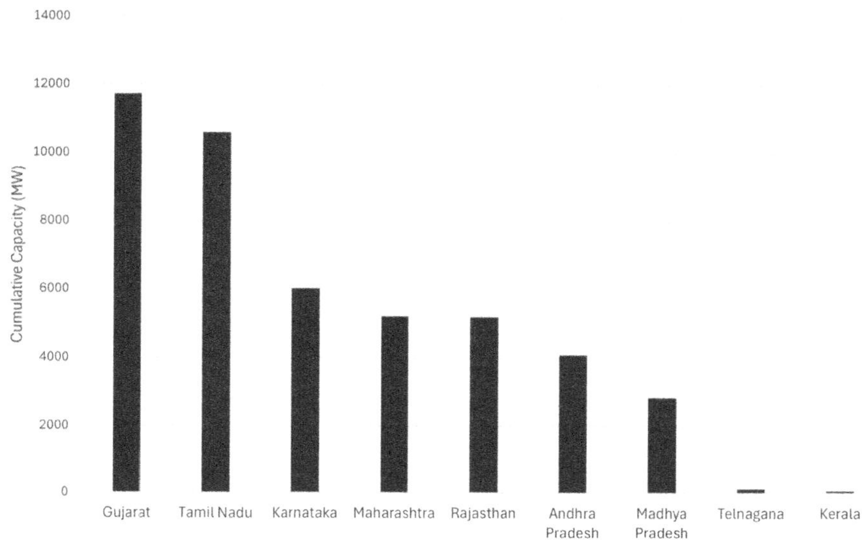

Figure 6.18 Indian states cumulative installed wind capacity as of 2024. MNRE (2025b).

potentials exceeding 70 GW combined (MNRE 2025a). Thus far, policy, cost, and technical barriers continue to hinder offshore wind development in a sustained manner.

6.5.2 Wind Energy Policy

The progression of wind-energy development in India is rooted in a policy-driven framework initiated in the 1990s. The Ministry of New and Renewable Energy (MNRE) institutionalized support and catalyzed rapid growth in wind-energy installations. From 2015 onward, India set ambitious targets to accelerate renewable energy integration, prominently featuring wind energy with goals of tripling the current installed capacity by 2030 (Koons 2024).

These aggressive targets have played a pivotal role in driving policy reforms and fostering local and foreign investment in wind infrastructure across the nation. More recently, the government has supported the development of a hybrid wind–solar–battery energy policy, recognizing the complementarity of such power sources in meeting diurnal and peak demands (Khare 2024; Shah 2024). It underscores a strategic approach to addressing solar and wind renewable energy intermittency. These hybrid systems could enhance grid stability and offer consistent renewable power, particularly during periods when solar output declines, and thereby mitigating reliance on conventional power sources (MNRE 2025c).

India's wind-energy policy framework comprises several pivotal regulatory measures outlined below, which are designed to foster rapid sectoral growth and attract investments (Sharma and Sinha 2019; Vaddadi 2023; Khare 2024; Shah 2024; MNRE 2025a). These measures are similar to what we see in other countries and include a varied mix designed to support the entire sector:

- Renewable Purchase Obligations (RPO)—Require state electricity distribution companies to source a specific percentage of their electricity from renewable energy sources. The MNRE sets increasing targets over time to encourage higher adoption of renewables.
- Inter-State Transmission System (ISTS) charges waived—To facilitate the interstate sale of wind power, the government waived transmission charges for wind and solar projects for a determined period. This policy reduces the cost burden on renewable energy producers and promotes the development of projects in high-resource regions, irrespective of the location of end-users.
- Competitive Bidding and Fiscal Incentives—Transparent competitive bidding processes introduced more recently have contributed to cost reductions, making wind power procurement economically attractive and competitive to conventional energy sources. The government also supports various tax breaks and accelerated depreciation benefits. India has offered targeted incentives, including tax exemptions and reduced customs duties on specific component parts. Moreover, supports to the domestic wind turbine assembly and manufacturing industry allow for localization multiplier effects through job creation and training.

6.5.3 India's Wind Farms

India's wind-energy landscape features several prominent wind-farm developments. Among those, the Jaisalmer Wind Park in the state of Rajasthan is developed by Suzlon Energy and recognized as one of the world's largest wind farms (Fig. 6.19). This facility crossed 1,000 MW installed capacity and comprises a network of smaller interconnected wind farms. Suzlon, a global wind power company with over 111 wind farms in India, engages in everything from initial feasibility studies to the operation and maintenance of wind-energy installations (Suzlon 2025). The company's comprehensive manufacturing and service infrastructure contributes significantly to India's renewable energy expansion.

The Muppandal Wind Farm in the state of Tamil Nadu is located in one of India's most wind-rich regions along the southeastern coastal belt (Shankar 2022). The Muppandal facility, advanced by the state's Energy Development Agency (TEDA), represents India's largest operational onshore wind farm and contributes

Figure 6.19 Bada bagh, a burial ground for royal families, in the foreground with the Jaisalmer Wind Park in the background. Truss towers of the Suzlon turbines have a decidedly industrial appearance. Original photo by Vineethasara (2017). Vineethasara / Wikimedia Commons / CC BY-SA 4.0.

significantly to Tamil Nadu's leading wind-energy status. Nearly 24% of India's wind-energy installed capacity comes from Tamil Nadu and the region around Muppandal, among others (MNRE 2025b).

The government's recent hybrid wind–solar energy policy has allowed for the development of the Hybrid Renewable Energy Park in the Kutch region of the state of Gujarat (Arasu 2023). It is planned and under construction as a 30 GW integrated wind–solar hybrid park. This initiative illustrates India's latest approach toward maximizing land efficiency and enhancing power reliability by combining complementary renewable technologies (Shah 2024).

6.5.4 Opportunities and Challenges

India's wind-energy sector stands at a pivotal juncture, offering substantial opportunities to bolster the nation's renewable energy portfolio. A strong and consistent policy framework has allowed the country to rise to a dominant fourth position globally in terms of installed capacity. Thus far, the capacity is predominantly distributed across the southern and western states, with Gujarat and Tamil Nadu leading in installations. Despite this progress, the sector faces challenges and opportunities that need addressing to sustain and accelerate growth.

Land acquisition for wind farms can be a complex and lengthy process especially when many landowners and vast tracks of land are involved. Administrative and regulatory hurdles which require extensive clearances from various governmental agencies may also prolong project implementation. Securing land for wind-energy projects has also emerged as a significant issue. In regions like Gujarat, Rajasthan, and elsewhere, local communities have protested wind turbine installations, citing loss of grazing land and environmental degradation (Langa 2021; Shankar 2022; Odeyemi 2024). These examples mirror those from across the globe where local communities have limited decision making input to large energy projects that affect their livelihoods. Such disputes highlight the tension between renewable energy development and traditional land use, emphasizing the need for more inclusive planning processes in approaches to decarbonization (Shankar 2022).

Consistent policies and regulatory certainties are a must for wind-energy investors and developers, as project investments and lifespans cross decades. Frequent changes in tariff structures and delays in approvals could deter investment and delay project implementation. A more stable and transparent policy environment is crucial for sustained growth and local and foreign investments in this sphere (Khare 2024).

The expansion of wind energy is also often hampered by inadequate transmission infrastructure (Gagal 2022). Bottlenecks in power transmission can lead to underutilization of installed capacity. For instance, delays in establishing transmission lines have impeded the pace of wind power expansion, underscoring the need for synchronized development of generation and transmission. The Indian government's recent planned investments into infrastructure improvements and transmission line expansions are examples to mitigate this challenge (Kwatra 2024).

The opportunities for wind energy in India to contribute to lowering carbon emissions from fossil fuels are promising (Lu et al. 2020). India's extensive coastline offers significant potential for offshore wind-energy development. The government has initiated policies to explore this avenue, which could substantially increase the country's renewable energy capacity (MNRE 2025a). Continuous improvements in turbine technology have enhanced energy capture and reduced costs. India's homegrown companies, such as Suzlon and Tata Power among others, have contributed to growth of the sector. The integration of wind energy with other renewable sources, such as solar, presents opportunities for hybrid systems that could provide more stable and efficient power generation. The Indian government's strategic planning in this arena is helpful (MNRE 2025c).

India's advancements in the wind-energy sector have allowed it to participate actively in the global renewable energy market. Exporting wind-energy technology and expertise opens new revenue streams. The wind-energy sector serves as a catalyst for economic development, generating employment opportunities across various segments, including manufacturing, installation, maintenance, and research. When done right, expanding this sector would stimulate economic growth and contribute to workforce development.

6.6 Summary

Eurasia is the world's largest landmass, contains a majority of the world's people, and has the greatest overall development of wind energy. The degree of wind-energy generation spans a broad range country by country from intensive development in some countries to negligible for many others. Denmark is a leader in green energy and has the highest per-capita wind-power capacity in the world. Emphasis during the twenty-first century has been for construction of larger wind farms, especially offshore. This includes two so-called energy islands, one an artificial island in the North Sea as well as Bornholm in the Baltic Sea. Many older, small-capacity turbines on land have been dismantled or replaced by fewer, larger, high-capacity turbines.

The Faroe Islands, a small cluster of islands in the North Atlantic, are a self-governing part of the Kingdom of Denmark. Like many relatively small and isolated islands, the Faroes are well positioned to take advantage of available renewable energy resources. In fact, the Faroes already produce more than half of the nation's electricity from renewable sources including hydropower, wind turbines, and tidal power. However, further development of hydropower and onshore wind turbines has met strong public resistance. A clear preference has emerged for future offshore development of renewable wind and tidal energy for the Faroe Islands.

Poland has emerged during the past three decades with a democratic government, strong economy, and membership in the EU and NATO. Poland experienced rapid growth in wind energy early in the twenty-first century, but this growth stalled after 2016 with strict rules for placement of new wind turbines. Russia' invasion of Ukraine in 2022 created a Polish energy crisis. Poland is in the process of building nuclear-power plants, and plans to spend heavily to develop offshore wind-energy resources in the Baltic Sea. The Port of Gdynia would serve as the land hub for a dozen offshore wind farms. Nonetheless, a full transition from coal and other fossil fuels to wind and nuclear energy would take at least two or more decades.

India ranks fourth overall for wind-generating capacity. India's extensive wind-energy potential varies significantly across different geographical regions due to diverse climatic and topographical conditions. The best wind resources lie in the country's western and southern portions. India has set ambitious targets to accelerate renewable energy integration supported by several policy and regulatory measures. Jaisalmer Wind Park and the Muppandal wind facility are among the largest, and the Hybrid Renewable Energy Park represents the move toward hybrid wind–solar–battery development. Nonetheless, securing land for new wind-energy projects has emerged as a significant problem with complex jurisdictional issues and local public resistance.

References

Arasu, S. 2023. In the salt deserts bordering Pakistan, India builds its largest renewable energy project. Associated Press, December 5, 2023. https://apnews.com/article/cop28-energy-transition-india-renwable-energy-project-climate-change-30d82470f77ed643d1dd949bfa263399

Barney, A., Petersen, U.R. and Polatidis, H. 2022. Energy scenarios for the Faroe Islands: A MCDA methodology including local social perspectives. *Sustainable Futures* 4. https://doi.org/10.1016/j.sftr.2022.100092

Barteczko, A. and Krasuski, K. 2023. Poland's next government halts wind energy plans after backlash. Bloomberg, December 5, 2023. https://www.bloomberg.com/news/articles/2023-12-05/poland-s-next-government-halts-wind-energy-plans-after-backlash

Berggreen, J. 2023. Another year, another record in Denmark's renewable energy progress. CleanTechnica. https://cleantechnica.com/2023/01/14/another-year-another-record-in-denmarks-renewable-energy-progress/

Cappelen, J. and Drost Jensen, C. 2021. Climatological standard normals 1991-2020 – Faroe Islands. The climate of the Faroe Islands – With climatological standard normals, 1991-2020. Danmarks Meteorologiske Institut. https://www.dmi.dk/fileadmin/Rapporter/2021/DMI_report_21_13_Faroe.pdf

Christensen, B. and Thorndahl, J. 2012. Fra husmøller til havmøller: Vindkraft i Danmark i 150 år. Energimuseet, Nordisk Folkecenter for Vedvarende Energi, Poul la Cour Foundation, Danmarks Vindkrafthistoriske Samling, 56 p.

CIA Poland 2025. Poland. The World Factbook, U.S. Central Intelligence Agency. https://www.cia.gov/the-world-factbook/countries/poland/

DEA 2015. The Danish energy model – Efficient and sustainable. Danish Energy Agency. https://stateofgreen.com/en/solutions/the-danish-energy-model-innovative-efficient-and-sustainable/

Ehlers, J. and Gibbard, P.L. 1996. Quaternary and glacial geology. John Wiley & Sons, Chichester, UK, 578 p.

Ember 2024. Per capita electricity generation from wind, 2023. Ember and Energy Institute – Statistical Review of World Energy (2024). Population based on various sources (2023) – With major processing by Our World in Data. https://ourworldindata.org/grapher/wind-electricity-per-capita

Energinet 2020. Preliminary studies for the world's first energy islands begin. https://en.energinet.dk/about-our-news/news/2020/11/30/energy-islands/

Energinet 2023. Energiøer i Danmark. https://energinet.dk/anlaegsprojekter/energioer/

Faroe Energy 2025. Energy. Official site of the Faroe Islands. https://www.faroeislands.fo/economy-business/energy/

Fedorska, A. 2023. Poland: Wind power runs out of steam. Deutsche Welle, February 2, 2023. https://www.dw.com/en/poland-wind-power-runs-out-of-steam/a-64666398

Gagal, V. 2022. How the weak transmission grid is an emerging challenge for the country's renewable energy projects. The Times of India, June 17, 2022. https://timesofindia.indiatimes.com/blogs/voices/how-the-weak-transmission-grid-is-an-emerging-challenge-for-the-countrys-renewable-energy-projects/

Geopark 2025. Discover Geopark Vestjylland. https://www.geoparkvestjylland.com/

Houmark-Nielsen, M. 2007. Extent and age of Middle and Late Pleistocene glaciations and periglacial episodes in southern Jylland, Denmark. *Geological Society of Denmark, Bulletin* 55, p. 9–35. https://doi.org/10.37570/bgsd-2007-55-02

IRENA 2024. Renewable energy statistics 2024. International Renewable Energy Agency, Abu Dhabi. ISBN ISBN: 978-92-9260-614-5. https://www.irena.org/-/media/Files/IRENA/Agency/Publication/2024/Jul/IRENA_Renewable_Energy_Statistics_2024.pdf

ITA 2021. Poland offshore wind terminal. U.S. International Trade Administration, Market Intelligence, April 14, 2021. https://www.trade.gov/market-intelligence/poland-offshore-wind-terminal

ITA 2023. Poland onshore wind energy 10H distance rule liberalized. U.S. International Trade Administration, Market Intelligence, April 18, 2023. https://www.trade.gov/market-intelligence/poland-onshore-wind-energy-10h-distance-rule-liberalized

Kardaś, S. 2023. From coal to consensus: Poland's energy transition and its European future. European Council on Foreign Relations, Policy Brief, September 27, 2023. https://ecfr.eu/publication/from-coal-to-consensus-polands-energy-transition-and-its-european-future/

Khare, D. 2024. Navigating the turbulent skies: Overcoming challenges and charting the future of wind energy in India. Power Line Magazine, May 8, 2024. https://powerline.net.in/2024/05/08/navigating-the-turbulent-skies/

Koons, E. 2024. Top wind energy companies in India 2024. EnergyTrackerASIA, January 31, 2024. https://energytracker.asia/wind-energy-companies-in-india/

Kwatra, P. 2024. Building a reliable network: Transmission expansion plans and projections up to 2031-32. Power Line, November 4, 2024. https://powerline.net.in/2024/11/04/building-a-reliable-network-transmission-expansion-plans-and-projections-up-to-2031-32/

Langa, M. 2021. Residents of Gujarat village protest against windmills being set up on forestland. The Hindu, August 7, 2021. https://www.thehindu.com/news/national/other-states/residents-of-gujarat-village-protest-against-windmills-being-set-up-on-forestland/article35781540.ece

Lu, T., Sherman, P., Chen, X., Chen, S., Lu, X. and McElroy, M. 2020. India's potential for integrating solar and on- and offshore wind power into its energy system. *Nature Communications* 11, p. 4750. https://doi.org/10.1038/s41467-020-18318-7

Marks, L. 2004. Pleistocene glacial limits in Poland. *Developments in Quaternary Sciences* 2, p. 295–300. https://doi.org/10.1016/S1571-0866(04)80079-4

Minesto 2025. The Faroe Islands. https://minesto.com/faroe-islands/

MNRE 2025a. Offshore wind. Ministry of New and Renewable Energy. https://mnre.gov.in/en/off-shore-wind/

MNRE 2025b. Renewable energy statistics, 2023-24. Ministry of New and Renewable Energy. https://mnre.gov.in/en/renewable-energy-statistics/

MNRE 2025c. Wind policies. Ministry of New and Renewable Energy. https://mnre.gov.in/en/wind-policy-and-guidelines/

NIWE 2023. Wind energy: Clean energy. Annual Report 2022–23. National Institute of Wind Energy, Ministry of New and Renewable Energy, Government of India. https://niwe.res.in/media/annual-reports/Annual_Report_2022-23_english_uXyZRqu.pdf

Nuclear Poland 2025. Nuclear power in Poland. World Nuclear Association. https://world-nuclear.org/information-library/country-profiles/countries-o-s/poland.aspx

Odeyemi, N. 2024. Wind turbines in India's Thar Desert stir controversy. The Electricity Hub, August 26, 2024. https://theelectricityhub.com/residents-in-indias-thar-desert-are-protesting-the-impact-of-wind-turbines-claiming-they-have-lost-grazing-land-and-suffered-from-damage/#:~:text=Residents%20in%20India's%20Thar%20Desert,power%20cuts%20and%20economic%20hardship

Pedersen, S.S., Salinas, I., Peterson, K.S. and Jensen, T.F. 1989. Jordartskort over Danmark 1:200.000: Midtjylland. Danmarks Geologiske Undersøglse, Copenhagen, Denmark.

Rasmussen, J. 1982. The Faeroe Islands: Geology. In, Rutherford, G.K. (ed.), The physical environment of the Faeroe Islands, Monographiae Biologicae 46, p. 15–34, Junk Publishers, The Hague.

SEV 2025. Framleiðsla beint nú. https://www.sev.fo/

Shah, R. 2024. Redirecting wind energy in India. Ember, October 8, 2024. https://ember-energy.org/latest-insights/redirecting-wind-in-india/

Shankar, P. 2022. The wind farm paradox in southern Tamil Nadu. Mongabay, April 15, 2022. https://india.mongabay.com/2022/04/the-wind-farm-paradox-in-southern-tamil-nadu/

Sharma, S. and Sinha, S. 2019. Indian wind energy & its development-policies-barriers: An overview. *Environmental and Sustainability Indicators* 1–2, 100003. https://doi.org/10.1016/j.indic.2019.100003

Strzelecki, M. and Evans, D. 2024. Poland may double wind power capacity by 2030, lobby says. Reuters, January 10, 2024. https://www.reuters.com/business/energy/poland-may-double-wind-power-capacity-by-2030-lobby-says-2024-01-10/

Strzelecki, M., Hardcastle, E. and Simao, P. 2023. Poland to ease wind farm rules, freeze energy prices. Reuters, November 29, 2023. https://www.reuters.com/business/energy/poland-ease-wind-farm-rules-freeze-energy-prices-2023-11-29/

Suzlon 2025. Group profile. https://www.suzlon.com/in-en/about-suzlon.corporate-profile

Thorndahl, J. 2011. En vindmølle i baghaven: Om vindkraft pioneren Christian Riisager. *Kapitler af Vindkraftens Historie i Danmark* 7, p. 2–8.

TPA 2025. Report wind energy in Poland 2024. TPA Poland. https://www.tpa-group.pl/publikationen/wind-energy-in-poland-2024/

Vaddadi, V.K. 2023. Wind energy in India – An overview. SAS Partners, October 30, 2023. https://saspartners.com/wind-energy-in-india-an-overview/

Wind Power 2025. The wind power: Wind energy market intelligence. https://www.thewindpower.net/media_maps_en.php

Wind Power Countries 2025. The wind power. Wind turbines and wind farms database: Countries list. http://www.thewindpower.net/country_list_en.php

Wind Power Margonin 2025. Margonin windfarm (Poland). http://www.thewindpower.net/windfarm_en_17117_margonin.php

Wind Power Poland 2025. Poland. The Wind Power. https://www.thewindpower.net/country_maps_en_27_poland.php

Wind Power Ramme 2025. Ramme (Denmark). The Wind Power: Wind turbines and wind farms database. https://www.thewindpower.net/windfarm_en_8472_ramme.php

Wind Power Udbyneder 2025. Udbyneder windfarm (Denmark). The Wind Power: Wind turbines and wind farms database. http://www.thewindpower.net/windfarm_en_7318_udbyneder.php

Wind-Turbine-Models 2025. Windturbines database: Enercon. https://en.wind-turbine-models.com/turbines?manufacturer=22&view=table

7

North American Case Studies

7.1 Introduction to the United States and Canada

Canada and the United States are the world's second and third largest countries, respectively, in land and sea area, spanning the continent of North America and reaching northward into the Arctic and westward across the Pacific Ocean. The United States is likewise the third largest country in population with >340 million people, but Canada is much smaller with less than 40 million people (CIA 2024). Both countries are rich in fossil fuels—oil, natural gas, and coal, and both countries have adopted wind energy as a priority for the future.

The two countries have quite different sources for generating electricity, however. The United States has an installed generating capacity of more than 1,200 GW. Of this capacity about 60% is fossil fuels and 18% is nuclear power. The remainder is generated by hydroelectric plants, geothermal and solar energy, and other renewable resources; wind accounts for about 10% of total generating capacity. Canada, on the other hand, has just 160 GW of installed generating capacity, of which more than 60% is hydropower and only 6% is wind power (CIA 2024).

The United States and Canada are self-sufficient in coal and, in fact, export substantial amounts of coal. Directional drilling and hydraulic fracturing (fracking) techniques have led to increased production of oil and natural gas. Canada exports significant petroleum mostly to the United States. Both countries export large volumes of natural gas (CIA 2024). This reliance on fossil fuels, particularly in the United States, results in substantial emissions of carbon dioxide. China is the world leader in this respect, more than twice the U.S. emissions. These two countries plus Canada together produce nearly one-half of all the CO_2 released as a consequence of burning fossil fuels (see Table 8.3).

Wind Energy Landscape: Principles, Techniques, History, First Edition. James S. Aber, Susan E. W. Aber, and Firooza Pavri.

Published 2026 by John Wiley & Sons, Inc.

Table 7.1 Top 12 U.S. states for installed wind capacity as of first quarter 2024.

State	GW	State	GW
Texas	37.2	Colorado	4.8
Oklahoma	11.8	Minnesota	4.2
Iowa	10.0	New Mexico	4.0
Kansas	8.8	Michigan	3.8
Illinois	7.7	North Dakota	3.7
California	6.2	South Dakota	3.5

Rounded values given in gigawatts (GW).
WindExchange (2024)/U.S Department of Energy.

The United States has been a world leader in wind energy since the 1980s. Early developments, particularly in California, showed the potential for electricity generation as well as the negative environmental and aesthetic aspects of uncontrolled development of wind farms (see Chap. 1 and 5). During the early twenty-first century, deployment of wind turbines and construction of wind farms has followed a more rational approach. Currently the United States ranks second behind China, and Canada is ninth for wind-power capacity (see Table 1.3).

Among individual U.S. states, 11 of the top 12 for installed wind capacity are located in the middle of the United States (Table 7.1). The exception is California, an early adopter of wind energy on the West Coast. This distribution reflects the primary wind-resource corridor of the central and interior western United States. As yet, offshore wind energy is still in its infancy, although New England has begun to develop some wind farms offshore from Rhode Island and Massachusetts. Long-term expansion of wind energy in the United States has spawned many programs in higher education for service training and engineering research.

Wind energy in Canada is spread across the country from Newfoundland to Inuvik in the Northwest Territories with a total of 330 wind farms (Wind Power Canada 2024). The greatest concentrations of wind farms are found in the Atlantic provinces, southern Quebec and Ontario, as well as the Canadian Prairie provinces of Saskatchewan and Alberta. In addition, offshore wind farms are planned for both the Atlantic and Pacific coastal regions.

In the United States, the Bipartisan Infrastructure Law (BIL) of 2021 and the Inflation Reduction Act (IRA) of 2022 provide substantial support for wind and solar power and related electricity infrastructure improvements. Specifically the BIL has

Wind-Power Training Programs

Across the United States, several dozen workforce training and education programs specialize in wind energy. Most of these programs are located in New England, the Midwest, and Great Plains regions, as well as southern California. Institutions range from major universities, to small technical and community colleges, to third-party training, and programs span the gamut from tower technicians to Ph.D. engineering degrees. The wind-turbine service technician is among the fastest growing jobs in the nation (WindExchange 2025).

A typical small example is the North American Wind Research and Training Center at Mesalands Community College in Tucumcari, northeastern New Mexico. The program offers a two-year Associate of Applied Science Degree in Wind Energy Technology as well as a Wind Energy Technology Certificate plus an Occupational Certificate in Basic Wind Energy Technology (MCC 2025). The curriculum includes both traditional course work and practical technical training dealing with turbine technology, construction, operation, maintenance, and safety as well as basic subjects in electricity, power generation and distribution, electronics, and communications.

Main building of the North American Wind Research and Training Center at Mesalands Community College, New Mexico. Note the stylized turbine-rotor windows. In the background stands a GE Wind 1.5 MW turbine, hub height 80 m, rotor diameter 77 m, installed in 2008. Hoen et al. (2018–2025). This turbine is used for hands-on training purposes.

wind initiatives both offshore and on land including technologies to minimize impacts to bats and local wildlife (DOE Infrastructure 2022). The IRA likewise addresses clean-energy production and storage with the goals of 950 million solar panels, 12,000 wind turbines, and 2,300 grid-scale battery plants (White House IRA 2022).

In Canada, federal and provincial governments offer a variety of financial and tax incentives for developing wind energy. Canada is now set, in fact, to become a leader in green-energy incentives with its renewable energy Investment Tax Credit (ITC) that was introduced into parliament in early 2024. In particular the Clean Electricity ITC would be available to Canadian corporations for investments in clean energy equipment (Der et al. 2024). Thus, the United States and Canada are poised for a major push to enhance wind energy.

7.2 Kansas

7.2.1 Introduction

Kansas has played a prominent role in the development of United States wind power. Like other Great Plains states, Kansas is well known for wind. Dodge City is the most windy city in the contiguous (48) United States, in fact. The first large wind farm came to Gray County in the High Plains of southwestern Kansas in 2001 (see Fig. 1.19), and wind energy expanded quickly across the state during the next two decades. The state currently ranks fourth in the nation for installed wind-generating capacity. More than 40% of total electric-generating capacity in Kansas is from wind energy, which is greater than combined coal plus natural gas capacity.

Wind energy has generated some $48 million in direct annual payments to Kansas landowners and $660 million in lifetime payments for local governments to support rural counties (Miller 2023). The Nature Conservancy has analyzed the wind-energy potential for Kansas along with impacts on wildlife and habitat. More than one-third of the state is suitable for wind energy based on engineering and land-use constraints, and nearly one-fifth of the state was identified as low impact for wildlife and habitat (TNC 2022). On this basis, approximately 40,000 km^2 would be available and well suited for further development of wind energy.

Clearly wind power has emerged in just two decades as a major energy resource for Kansas. As of mid-2024, Kansas had more than 4,200 wind turbines with a total generating capacity of 9.1 GW (Hoen et al. 2018–2025). In addition to these are new wind farms under construction, plus individual smaller wind turbines, as well as arrays of solar PV panels. The following assessment and review of the Kansas windscape is based on two decades of observations that are summarized in Aber and Aber (2024) as well as other cited references.

Small Turbines, Kansas

In addition to the large wind turbines seen throughout Kansas are smaller turbines scattered around the state. Some of these are included in the U.S. Wind Turbine Database, for example, three small Nordtank and Northern Power turbines operated by Cloud County Community College and Cloud Power in conjunction with a solar farm. However, many of the smaller turbines are not included (Hoen et al. 2018–2025). Such smaller turbines serve individual farms, ranches, and business enterprises.

These turbines may be new, used, or remanufactured equipment. Some are built in the United States. Others came originally from Denmark supplied by Bonus, Nordtank, and other companies. The Danish turbines were delivered to California in the 1980s, then later dismantled, refurbished, and resold across the country. For the most part, these small turbines have generating capacities less than 100 kW.

(A)

(B)

Small wind turbines manufactured in the United States and operating in Kansas. (A) Excel 10 kW turbine combines with solar panels to provide electricity for a building on the Kansas State University Salina campus. Excel is a product of Bergey Windpower from Oklahoma, (B) Northern Power Systems (NPS) turbine at a business in Rush Center. NPS originated in 1974 and is based in Vermont.

7.2.2 Kansas Windscape

In general, the western three-quarters of Kansas have the best wind resources with annual average wind speed of 7.0 m/s or greater at 50-m height. Contrary to popular opinion, Kansas is not flat; the state actually has considerable elevation relief, which is important for siting wind farms. The most favorable localities are high topographic locations on drainage divides and prominent bedrock escarpments, particularly in the High Plains, Blue Hills, and Flint Hills, where average wind speed exceeds 7.5 m/s at 50-m height (Fig. 7.1).

Nearly all wind farms in Kansas, in fact, are situated on drainage divides (Fig. 7.2). The placement of wind farms on drainage divides reflects patterns created by long-term stream erosion since the end of the Miocene Epoch about five million years ago (Aber et al. 2023). Overall, the distribution of wind farms across Kansas exploits these patterns of erosion and the resulting drainage divides (see sect. 3.1.4). Most development of wind energy in Kansas has taken place in the High Plains and Blue Hills, and selected wind farms are described below. The Flint Hills, however, has seen only limited development for wind energy.

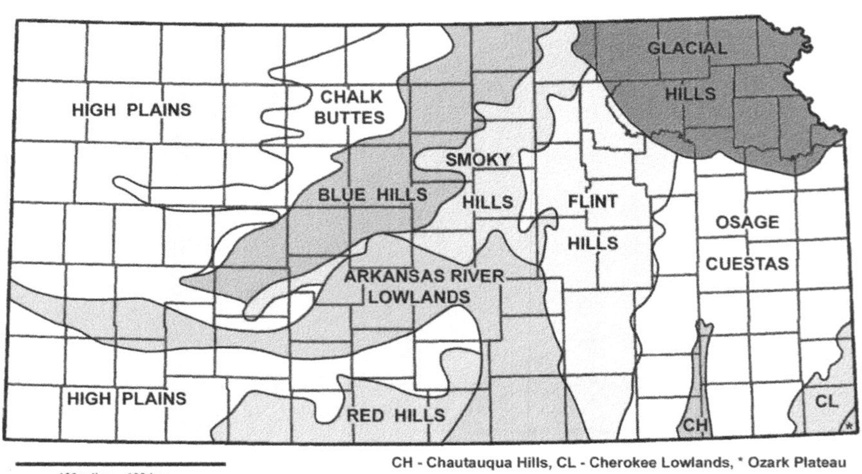

Figure 7.1 Landscape regions of Kansas. Region limits are sharp and well defined in some cases, but may be transitional or gradual in other places. County boundaries shown in gray. Aber et al. (2023).

Figure 7.2 Marshall Wind Energy complex in the Glacial Hills of northeastern Kansas. Vestas V110, 2.0 MW turbines; total height is 150 m. The turbines are located on a local drainage divide, known as the Summit, that exceeds 440 m elevation. This wind farm began operation in 2016. Hoen et al. (2018–2025).

The Meridian Way Project and Project II wind farms are located on the Blue Hills escarpment in Cloud County, north-central Kansas. This escarpment is supported by the Greenhorn Limestone (late Cretaceous age) and stands more than 50 m above lower terrain to the south. This position is also on the drainage divide between the Solomon River and Republican River basins, and so intercepts winds from all directions. Sixty-seven Vestas V90 3.0 MW turbines were erected in 2008 and have a total rated capacity of 201 MW (Hoen et al. 2018–2025). The turbines are deployed in two arrays to the west and east of U.S. highway 81 along the drainage divide (Fig. 7.3). Total height of the turbines is 125 m.

The wind farm is entirely compatible with mixed agricultural land use and has proven beneficial for some wildlife (Fig. 7.4). Among the more important environmental indicators for optimal grassland habitat are greater and lesser prairie-chickens (*Tympanuchus cupido* and *T. pallidicinctus,* respectively). Multi-year investigation has shown surprisingly that prairie-chickens are not adversely affected by wind farms; in fact, female survival rates increased after wind turbines were installed at the Meridian Way wind farm (Sandercock 2013).

Tallgrass Heartland, Kansas

In 2004, Kansas Governor Kathleen Sebelius proposed the *Heart of the Flint Hills Area*, in which wind farm development would be discouraged through voluntary restraint in order to protect the tallgrass-prairie ecosystem. The governor's decision followed intense debate involving land owners, local governments, energy developers, and numerous environmental organizations. In some Flint Hills counties, already local officials had banned the construction of large wind farms, because of the perceived negative social, cultural, and aesthetic impacts. Conversely other counties favored the economic development and tax base that wind farms would bring.

The original moratorium area was relatively restricted in geographic coverage to that portion of the Flint Hills that preserved the most intact tallgrass habitat, was least altered by human activities, and had the greatest scenic beauty. Elk River and Caney River wind farms are situated to the south, outside the original Heart of the Flint Hills Area. In 2011, Governor Brownback more than doubled the extent of the exclusion zone, enclosing a region approximately 100 km east–west by 275 km north–south, which he renamed

Kite aerial photographs of wind farms in tallgrass prairie of the southern Flint Hills. (A) Elk River in southeastern Butler County (2005). Total of 100 GE Wind turbines with a combined capacity of 150 MW. (B) Caney River in western Elk County (2011). Total of 111 Vestas V90 turbines with a combined capacity of 200 MW. Hoen et al. (2018–2025).

Tallgrass Heartland, Kansas (Continued)

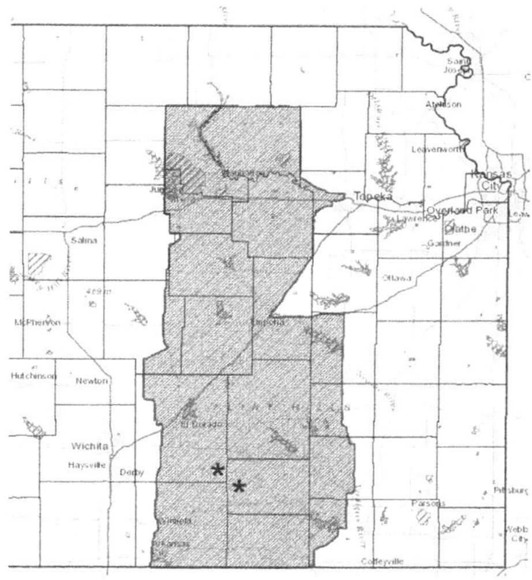

Eastern Kansas showing the extent of the Tallgrass Heartland (diagonal lining), as established by Governor Brownback (2011) and reaffirmed by Governor Kelly (2020). Elk River and Caney River wind farms are marked (*). Aber et al. (2015)/Elsevier.

the *Tallgrass Heartland*. The expanded region includes most of the Flint Hills as well as parts of adjacent regions.

Brownback's decision was announced as an attempt to balance economic development, energy resources, and environmental protection. Within the expanded exclusion zone, Elk River and Caney River wind farms were allowed to continue, and construction of new transmission lines across the Flint Hills is also allowed in order to transport wind-generated electricity from numerous large wind projects in central and western Kansas. Governor Kelly in 2020 reaffirmed the Tallgrass Heartland wind moratorium, marking the third political administration to protect this unique tallgrass-prairie ecosystem (Kansas Governor 2020).

In the meantime, the U.S. Fish and Wildlife Service and the Nature Conservancy along with other partners created the Flint Hills Legacy Conservation Area (FHLCA). The targeted region corresponds closely with Selelius' original moratorium area. The primary means of habitat protection are perpetual conservation easements with private land owners with an emphasis on grassland and wetland habitats. Such easements typically prohibit future residential and commercial development, drilling and mining, plowing, or other soil disturbances (Obermeyer et al. 2011). However, traditional land uses, such as livestock grazing and haying, would be permitted. The goal is to maintain the tallgrass-prairie habitat, while allowing historical types of agricultural practices to continue. As of 2023, more than 100,000 acres (>40,000 ha) have been protected within the FHLCA (TNC 2023).

Figure 7.3 Eastern end of the Meridian Way Project. Note the weather tower, which is a feature of all wind farms for monitoring wind and other atmospheric conditions. A transmission line crosses the foreground. Kite aerial photograph (2013).

Figure 7.4 Courtship display by greater prairie-chickens (*Tympanuchus cupido*) on a private ranch in Kansas. USFWS Mountain-Prairie/ Wikimedia Commons/CC BY 2.0.

The Flat Ridge Wind Farm was established in 2009 in Barber County, south-central Kansas. It had originally 40 Clipper C96 turbines with 2.5 MW capacity; some of these were replaced later with Vestas V110 turbines with 2.0 MW capacity. The array extends approximately east–west in multiple lines on a narrow, mesa-like finger of the High Plains that forms the drainage divide, some 90 m above the Medicine Lodge River basin to the south. Many turbines are situated in cattle pastures to minimize disturbance of the crop fields (Fig. 7.5).

This vicinity is underlain by the High Plains Aquifer in the Ogallala Formation, and groundwater from the aquifer supports irrigated agriculture. Turbines are placed so as not to interfere with center-pivot irrigation systems (Fig. 7.6). Since the original wind farm was built, several more have been erected, including Flat Ridge 2 and Flat Ridge 2 Expansion, both with GE Wind turbines in 2012. The final addition is Flat Ridge 3 in 2021. The latter includes the largest turbines, Vestas V120, rotor diameter 120 m, total height 152 m, capacity 2.2 MW. The total Flat Ridge energy complex contains 395 turbines with a combined capacity of nearly 700 MW (Hoen et al. 2018–2025).

Figure 7.5 Flat Ridge Wind Farm. Original Clipper C96 turbines stand 128 m tall with a blade in the upright position. Two closest turbines are spaced about 300 m apart. Note the mixed land use with crops (green winter wheat) and grazing cattle (black dots) on prairie grass in this early spring view looking westward. Kite aerial photograph (2011).

Figure 7.6 Clipper C96 turbines of the Flat Ridge Wind Farm with a center-pivot irrigation system in the foreground. Helium-blimp aerial photograph (2011).

The Flat Ridge vicinity demonstrates some typical characteristics for wind energy across the state. Locations that are favorable tend to grow through time as additional turbines and wind farms are added along with substations and transmission lines (Fig. 7.7). Furthermore, the size of turbines has increased

Figure 7.7 Electricity grid infrastructure under construction in Kansas. Transmission line in connection with the Sunflower Wind farm near Florence (2023).

through time in terms of height, rotor diameter, and rated capacity (Table 7.2). These trends reflect developments throughout the global wind industry.

7.2.3 Kansas Energy Resources

Coal, petroleum (oil), natural gas, and nuclear (uranium) were primary fuel sources for generating Kansas electricity in the twentieth century. Kansas was once a major oil-producing state; in fact, it ranked third in the nation behind California and Oklahoma in 1918 (Fig. 7.8). But this status has slipped over the past century, and Kansas accounted for only 1% of national oil production in 2022 (EIA Kansas 2024).

Kansas also was formerly a major source for coal in the late nineteenth and early twentieth centuries. However, most coal mines ceased operating in the mid-twentieth century, and the last small coal mine in Kansas shut down in 2016. Coal used in Kansas power plants today is shipped from Wyoming's Powder River Basin (Fig. 7.9). Numerous sources of natural gas are found in Kansas, including a sizable portion of the huge Hugoton Gas Area, which is among the largest gas

Table 7.2 Growth of typical turbine characteristics in Kansas wind farms during the early twenty-first century (2001–2021).

Year	Wind farm	Tower height	Rotor diameter	Maximum height	Spacing	Capacity (MW)
2001	Gray County	65	47	88	160	0.66
2005	Elk River	80	77	118	215	1.5
2009	Flat Ridge	80	96	128	300	2.5
2012	Flat Ridge 2	80	100	130	320	1.6
2017	Bloom	92	117	150	400	3.3
2019	Prairie Queen	114	132	180	500	3.55
2021	Iron Star	108	149	182.5	500	4.8

Hoen et al. (2018–2025).

Figure 7.8 Two working pump jacks and oil storage tank in the El Dorado Oil Field. It was discovered in 1915, and by 1918 it was the largest single oil field in the United States. It remains in production today albeit at a much-reduced rate. The power lines provide electricity to run the pumps.

fields in the world. As a relatively clean fuel in abundant supply, electric utilities have turned increasingly to natural gas. Nonetheless, the state consumes about twice as much gas as it produces (EIA Kansas 2024).

Kansas has one nuclear power plant that went online in 1985 and is licensed to operate until 2045. However, the allure of nuclear energy has faded with concerns about safety, mining, proliferation, disposal of nuclear wastes, and high cost. Meanwhile, wind energy has grown dramatically during the past quarter century.

Figure 7.9 BNSF Railway coal train at Las Animas in southeastern Colorado heading east toward the coal-fired electric generating station at Holcomb, Kansas, and perhaps beyond.

In 2022, nearly half of the state's electricity net generation came from wind turbines, which was the third highest among states (EIA Kansas 2024).

It is clear that a major shift in energy production and consumption is well underway at the state level. Nonetheless, Kansas remains a major center for oil refining, ethanol production, natural-gas storage, and pipeline transmission of fossil fuels. Agriculture, railroads, construction, and truck transportation are still heavily dependent on diesel fuel. Electric-vehicle charging stations have begun to appear around the state, although many rural areas remain poorly served. Thus, the green-energy transformation has a long way to go in Kansas.

7.3 Southeastern Colorado

7.3.1 Colorado Windscape

Colorado had more than 4.8 GW of installed wind-generating capacity as of first-quarter 2024, which ranks seventh among states (see Table 7.1). Wind farms are located almost entirely on the thinly settled High Plains east of the Rocky Mountains. Wind farms exploit uplands and drainage divides in which average annual wind speed generally exceeds 7.0 m/s at 100 m height (WindExchange States 2024).

South of the Arkansas River valley, the High Plains of southeastern Colorado host 10 wind farms that range in age from 2003 to 2019 (Hoen et al. 2018–2025). Four small wind farms each have just one to four turbines, and six larger wind farms vary from 16 to more than 100 turbines each. Altogether, 274 turbines have a total rated capacity of 477 MW. The following discussion focuses on a cluster of three large wind farms located along Colorado highway 10 about 30 km east of Walsenburg.

Huerfano River, Colorado

This is a typical example of a small wind farm near Walsenburg in south-central Colorado. The San Isabel Electric Association put up its first temporary, 30-m-tall meteorological tower in Huerfano County in 2004 to collect wind data near the westernmost edge of the High Plains at an elevation about 1,800 m (SIEC 2014). Colorado law required that electric cooperatives must by 2013 supply 20% of their power from renewable resources, not to include federal hydropower.

Originally planned for five turbines, construction started in the summer of 2013. Four Sany Electric turbines were erected next to an existing substation that connects to adjacent transmission lines along the I-25 highway corridor. The tower manufacturer was German-based REpower. The Chinese turbines stand 143 m in total height and have 80-m-diameter rotors with 2.0 MW nominal output for a total capacity of 8 MW (Hoen et al. 2018–2025). This represented 10% of San Isabel's maximum annual peak demand at that time (SIEC 2014). The $17 million project began routine operation late in 2013 and continues in service more than a decade later.

Temporary meteorological tower erected north of Walsenburg, Colorado, for the Huerfano River Project, as seen in 2011. Typical spring thunderstorm moves across the High Plains in the background.

Huerfano River Project completed as seen in 2014. Huerfano River substation operated by the San Isabel Electric Association with turbines visible in the background. Sany Electric turbines, model SE10020IIIE.

7.3.2 Cordova Mesa and Rattlesnake Buttes

Peak View Wind farm is situated on the eastern edge of Cordova Mesa (Fig. 7.10), and nearby the Busch Ranch and Busch Ranch II wind farms are located at North and South Rattlesnake Buttes, also known as Twin Buttes. Peak View Wind began operation in 2016 and is equipped with 34 GE Wind turbines that stand 130 m tall. Total rated generating capacity is 61 MW. The Busch Ranch wind farms include 43 Vestas turbines of 130 and 140 m height (Fig. 7.11). These began operating in 2012 and 2019 with a combined rated capacity of 88 MW (Hoen et al. 2018–2025). The situation of these three wind farms is a good example of where the underlying geologic structure is clearly expressed in surface topography.

Cordova Mesa is located on the western flank of the Apishapa anticline, a structure in which rock layers are domed upward into a broad arch. An escarpment capped by resistant Cretaceous limestone forms the mesa and surrounds the oval-shaped anticline (Stose 1912). The crest of the anticline is marked by Rattlesnake Buttes, which also are capped by the same limestone (Fig. 7.12). Older and more erodible sandstone and shale are exposed in the lower center of the uplift. In addition, younger igneous dikes have intruded across the anticline and stand as ridges in the landscape. The most prominent of these is the Hogback that rises more than 30 m above surrounding terrain in the Busch Ranch II wind farm.

Figure 7.10 GE Wind turbines in the Peak View Wind farm on the edge of Cordova Mesa with a traditional Aermotor windmill for pumping groundwater in the lower foreground.

Figure 7.11 Vestas V100 turbine under construction for the Busch Ranch wind farm in 2012 at North Rattlesnake Butte. Rotor diameter is 100 m, hub height 80 m, and total height 130 m.

Figure 7.12 North Rattlesnake Butte (left) stands in the middle of Busch Ranch wind farm and is capped by resistant Fort Hays Limestone of late Cretaceous age. Stose (1912).

Figure 7.13 Transmission line in the Busch Ranch wind farm collects electricity from substations of all three wind farms and sends the electricity north toward Pueblo, Colorado.

The mesa, buttes, and hogback mark the drainage divide between the Cuchara River and Apishapa River basins. North and South Rattlesnake Buttes exceed 1,950 m elevation, and the crest of Cordova Mesa reaches 1,980 m, some 120 to 150 m above adjacent lower terrain (NGMD 2023). The turbines take advantage of this conspicuous topographic setting for the greatest potential wind energy. Average annual wind speed at 100 m height is 8.0–9.0 m/s (WindExchange States 2024).

Major electricity use in Colorado takes place in several large urban areas, where the High Plains meet the Rocky Mountains. Each of the three wind farms has a substation that connects with a transmission line that runs northward to other transmission lines that serve the San Isabel Electric Association based in Pueblo (Fig. 7.13; Energy Atlas 2025). Installed generating capacity of the three wind farms totals 149 MW, which is sufficient for about 40,000 homes. To put this in perspective, Huerfano County and Las Animas County, where the wind farms are located, together have some 11,200 housing units (Census Bureau 2022).

7.3.3 Environmental and Aesthetic Issues

Visibility of wind turbines is a factor that some people find displeasing. Given their large size and topographic prominence, these turbines are visible from a long distance away. On a clear day, in fact, the turbines may be seen all the way to the Cuchara Valley

Airport at La Veta, a distance of some 55 km to the west. However, visibility of turbines is no more disturbing than numerous radio and cell-phone towers in the same vicinity.

Human population is quite sparse in the wind-farm neighborhood. Nonetheless, a few widely scattered homesteads are present in the vicinity (Fig. 7.14). Cattle ranching is the primary land use, and impact of turbines on cattle and horses seems trivial. Likewise impact on wildlife appears to be minimal as mule deer (*Odocoileus hemionus*) and pronghorn (*Antilocapra americana*) are quite common within and around the wind farms (Fig. 7.15).

Figure 7.14 Turbines appear to be quite close behind this farmstead on the Busch Ranch, Huerfano County Road 130. In fact, the nearest turbine (on right) is more than 1 km beyond the farmstead. Vestas V120, 2.2 MW turbines, rotor diameter 120 m, hub height 80 m, total height 140 m. Hoen et al. (2018–2025).

Figure 7.15 Pronghorn (*Antilocapra americana*) are commonly seen on the High Plains in the vicinity of wind farms at Rattlesnake Buttes and Cordova Mesa.

7.4 Offshore New England

7.4.1 Offshore Windscape

During the past decade, the coastal states of New England, including Maine, Massachusetts, Rhode Island, and Connecticut, have worked to advance their offshore wind energy sector. This movement has been driven by a commitment to increase the mix of renewable sources of energy to help meet state and regional carbon emissions reduction mandates, enhance energy security, and support the development of the Blue Economy, which refers to all economic activities related to oceans, seas, and coasts (EU 2021). Even so, offshore wind energy development is in its infancy in the United States when compared to China or European countries like the United Kingdom, Denmark, Germany, Belgium and others, who have substantially expanded offshore production (IRENA 2024).

Coastal New England exhibits significant potential for offshore wind energy development, with its oceanographic characteristics offering both opportunities and challenges. The northern section, encompassing the Gulf of Maine along the coasts of Maine and New Hampshire, is characterized by a rugged and broken coastline with relatively deep waters short distances from the shore. Depths are between 100 and 300 m across the Gulf's most suitable wind energy sites (Musial et al. 2023). The deep waters across the Gulf of Maine have made it more suitable for floating-turbine technologies. Offshore wind resources vary seasonally but are among the more robust found along the U.S. east coast, generally >9 m/s at 100 m height (Fig. 7.16). Although hurricanes are less of a concern to offshore

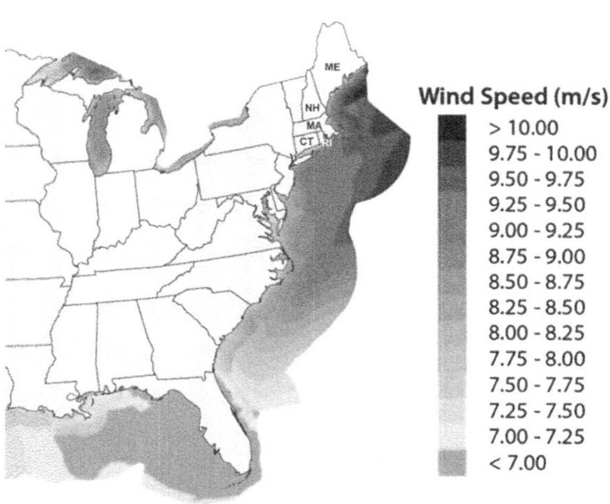

Figure 7.16 Offshore wind-energy resource for the eastern United States. Average wind speed at 100 m height. New England coastal states: ME—Maine, NH—New Hampshire, MA—Massachusetts, RI—Rhode Island, and CT—Connecticut. Adapted from Hartman (2016)/U.S Department of Energy/Public Domain.

turbine sites this far north in latitude, nor'easters are a common occurrence. Turbine technologies have been developed to withstand the impacts of extreme weather events.

The southern portion of New England, including the coasts of Massachusetts, Rhode Island, and Connecticut, features a more varied seabed topography with a somewhat gently sloping shelf into deeper waters (Ward et al. 2020). Nearshore water depths are generally shallower than in the Gulf of Maine, and consequently traditional fixed-bottom offshore wind turbines are the norm. The wind resources are similarly robust and appropriate for energy production, with distances farther from the shore providing more suitable conditions. Collectively, the geography of the New England coast, the maritime economies, as well as favorable policies have drawn significant investment interest in recent years.

7.4.2 Offshore Capacity and Technology

Offshore wind energy production capacity in the United States stood at 103 GWh in 2016, which increased to 145 GWh by 2023 (IRENA 2024). In comparison, China's offshore production capacity was 63,572 GWh, the United Kingdom was at 45,020 GWh, and Germany stood at 25,124 GWh in 2023. Like onshore wind energy installations, the timeline from the initial planning stages to site control, permitting, approvals, financial closings, construction, and finally operability and grid connectivity can be substantial, involving local-, state-, and federal-level approvals (McCoy et al. 2024). Added to these, policy signals governing offshore wind development are critically important. New England states have all implemented policies favorable to the development of offshore wind energy. However, President Trump's 2025 Executive Order temporarily pausing federal approval of all offshore wind leases and permits has introduced uncertainty and heightened risk to the sector (Shankman 2025).

Offshore wind turbines are specifically designed to withstand the harsh marine environment. In the United States, the primary technology used is fixing turbines to the ocean floor in 60 m or less of water depth. Offshore turbines tend to be larger in size due to fewer obstructions and hence have power-generating capacities that may be double that of their land-based counterparts (Fig. 7.17). Their engineering is more complex to accommodate fatigue and corrosion, rough ocean conditions, strong winds and large waves, among others. Floating deep-ocean turbine technologies in the United States are still in their developmental phase, with the University of Maine experimenting with a patented concrete-hull floating offshore wind turbine for deep-water deployment (UMASCC 2025). Given the deep waters of the Gulf of Maine, traditional fixed-bottom turbines are not feasible.

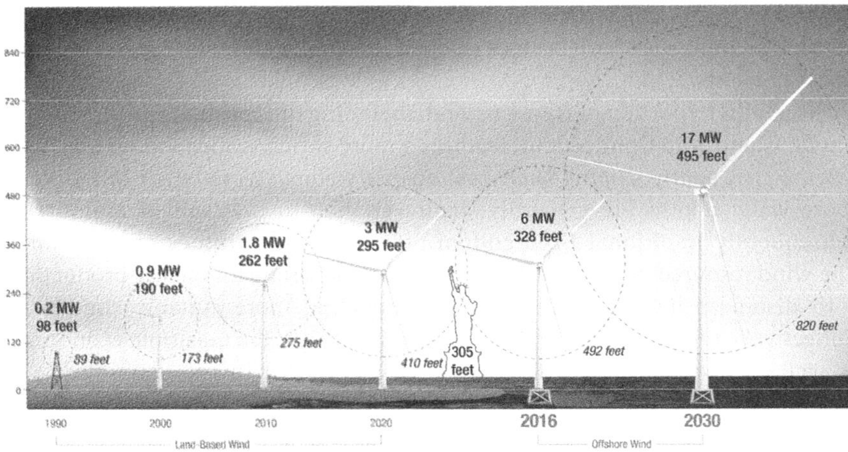

Figure 7.17 Increasing size and generating capacity of land-based and offshore turbines through time. Hub-height and rotor-diameter measurements given in feet. Statue of Liberty is shown for comparison (305 feet = 93 m). WindExchange Offshore (2024)/U.S. Department of Energy/Public Domain.

7.4.3 Offshore Wind Development Trajectory

New England's exploration of offshore wind energy was initiated with the Cape Wind project in 2001, which proposed to install a total of 130 turbines just roughly 8 km from the Massachusetts coast and near the Cape Cod, Nantucket, and Martha's Vineyard region. This project was also the United States' first foray into the offshore wind energy space. Despite its potential to generate 460 megawatts (MW) of electricity, the Cape Wind project faced extensive regulatory hurdles and strong local opposition with multiple lawsuits filed, leading to its eventual cancellation in 2016 (Davidson 2018).

Despite the troubles with New England's first proposed offshore project, state governments began to recognize the potential of wind energy. The period from 2015 to 2024 witnessed significant policy advancements at both state and federal levels, fostering the growth of on-land and offshore wind energy in New England. States initiated extensive studies to assess wind resources, invested in technologies, and explored legislative avenues to promote wind and renewable energy integration.

Marking a significant milestone, Rhode Island-based developer Deepwater Wind, later bought out by Ørsted, installed the U.S.'s first operational offshore wind farm, the 30 MW Block Island Project in 2016 (Fig. 7.18). Since this pioneering effort, Rhode Island has committed to a request for proposals to expand offshore capacity by 600 to 1,000 MW with the aim of potentially meeting 30% of the state's

Figure 7.18 Aerial view of the Block Island Project that includes five GE Wind turbines with a total capacity of 30 MW. Located in coastal Rhode Island, the project went online in 2016 and was the first operational offshore wind farm in the United States. Total height of turbines in 181 m (594 feet). NOAA (2023)/NOAA/Public Domain.

estimated 2030 electricity demand, equivalent to powering approximately 340,000 homes annually (Whitty 2022).

Connecticut has demonstrated a strong commitment to integrating offshore wind energy into its renewable energy portfolio. Efforts have focused on a sustained long-term strategy to expand the state's renewable portfolio and fully decarbonize the electricity sector by 2040 (DECD 2023). Its strategy focuses on leveraging the region's historical maritime strengths to drive workforce development in the blue economy, enhance deep-water port terminals and infrastructure to support the growing offshore wind industry, and strengthen grid integration. This includes public–private investments in key locations such as New London, Bridgeport, and New Haven aimed at expanding turbine assembly, staging, and deployment areas for future projects. Currently Revolution Wind, a 704 MW project with operations based out of New London, will include approximately 65 turbines between Block Island, Rhode Island, and Martha's Vineyard, Massachusetts, and is expected to be operational by 2026 (Fitzpatrick 2024).

Massachusetts has been at the forefront of offshore wind energy development in the United States. The state has implemented favorable policies including the Driving Clean Energy and Offshore Wind Act in 2022, which represents a significant legislative effort to position that state as a leader in offshore wind development. Components of this legislation include increasing the offshore wind procurement target to 5,600 MW by 2027, providing tax and other incentives to localize manufacturing and allied industry and workforce developments, removing price caps, and supporting improvements to the transmission infrastructure (CoM 2022).

Massachusetts' Vineyard Wind 1 project south of Martha's Vineyard started contributing to the grid in 2024 with five turbines providing 68 MW of power. It is Spanish firm Iberdrola's and subsidiary Avangrid Renewables' first offshore project in the United States. The project proposes 62 turbines and 800 MW of wind energy powering up to 400,000 homes when fully completed and deployed (Iberdrola 2025). Additional projects in Massachusetts include SouthCoast Wind with 804 MW capacity slated to come online by 2027 and Commonwealth Wind with 1,232 MW and completion scheduled for 2028 (NEOW 2025).

The state of Maine has recognized the offshore wind energy potential of the Gulf of Maine and has implemented strategic steps to harness this resource. The Maine Offshore Wind Initiative has explored opportunities and developed the Maine Offshore Wind Roadmap, a comprehensive plan outlining strategies for port infrastructure investments, supply-chain development, localized research and development involving universities, and stakeholder engagement (GEO 2023). In 2023, Maine passed the Procurement of Energy from Offshore Wind Resources Act advancing a framework for offshore wind energy development. It sets a target of 3,000 MW from offshore wind energy fed into the electricity supply for state residents by 2040. Further, it supports the development of ports through public–private partnerships and commits developers to investments in workforce training and local community benefits.

The state has committed to investing in a port on Sears Island, Maine to support the fabrication, staging, assembly, and maintenance of future offshore wind projects (GEO 2023). Emblematic of other similar instances, this site's selection has prompted mixed reactions from local communities and environmental organizations. While some stakeholders support the project for its potential economic benefits, others express concerns about the development, environmental impact, and preservation of the island's natural condition, particularly since an alternate site at close-by Mack Point already houses some of the needed port infrastructure (Ropeik 2024). In October 2024, however, the federal government declined the state's application for $456 million in funding to construct the offshore wind port at Sears Island (Singer 2024). This decision as well as a new federal policy impacting wind energy from the Trump administration pose challenges to the project's financing and timeline.

The Gulf of Maine's deep waters make floating rather than fixed turbine technologies a necessity. The University of Maine's Advanced Structures and Composites Center developed the VolturnUS, a floating concrete-hull technology designed for deepwater deployments (UMASCC 2025). This innovation has positioned Maine as a potential leader in floating offshore wind technologies, offering solutions adaptable to various marine environments. In 2024, the federal Bureau of

Ocean Energy Management (BOEM) agreed to the state's request for a floating offshore wind research lease. The proposed research array project is a planned 12-floating-turbine offshore wind pilot project, located 30 miles off the state's coast, and aims to demonstrate the viability of floating wind turbines in deepwater settings. The project will use the University of Maine's floating platform technology and partner with Diamond Offshore Wind to deploy the turbines (Office of Governor 2024).

7.5 Saskatchewan, Canada

7.5.1 Overview

In general terms, the southern one-third of Saskatchewan is an open, wind-swept, prairie landscape (Padbury and Acton 1994). Most of the landscape is utilized for crop agriculture and cattle ranching, which has led to Saskatchewan's reputation as the "Bread basket of Canada." This is also where the majority of people live and consumption of electricity takes place, particularly in the larger cities of Regina and Saskatoon.

Eleven wind farms are found in southern Saskatchewan, most located along or south of the Trans-Canada Highway 1. Total generating capacity is approximately 620 MW (Wind Power Canada 2024). Of these wind farms, the majority is situated in the Alberta Plain (Fig. 7.19). The northeastern margin of the Alberta Plain is marked by the Missouri Coteau, which forms the continental drainage divide

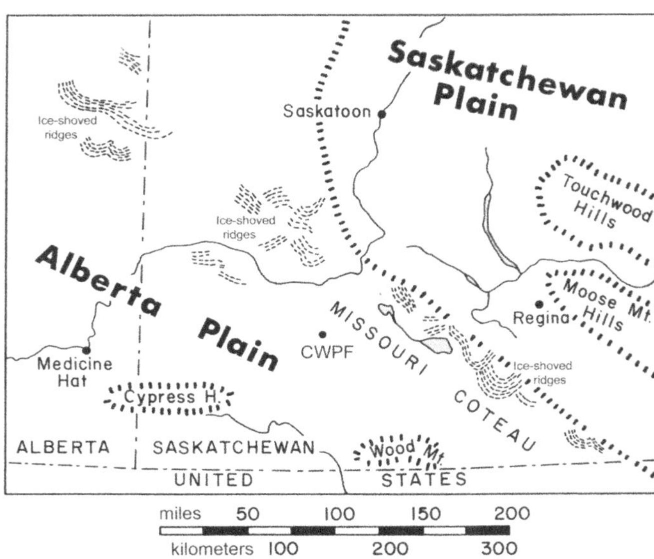

Figure 7.19 Sketch map of southwestern Saskatchewan and adjacent areas. Most wind farms are located in the Alberta Plain. A series of large ice-shoved ridges follow the crest of Missouri Coteau, which marks a major drainage divide between the Saskatchewan River basin to the north and the Missouri River basin to the south. Approximate location of the Centennial Wind Power Facility (CWPF). Aber and Ber (2007)/Elsevier.

between the Missouri–Mississippi basin to the south and the Saskatchewan–Hudson watershed to the north. The coteau is a hilly glacial upland standing about 150 m above the lower Saskatchewan Plain to the northeast.

7.5.2 Centennial Wind Power Facility

The Centennial Wind Power Facility is located on the Missouri Coteau about 25 km southeast of the city of Swift Current. Elevation of the wind-farm vicinity is approximately 880 m, some 140 m above Swift Current. The Centennial wind farm was built in 2005 and commissioned in 2006. The name comes from Saskatchewan's centennial in 2005. At the time it was commissioned, it was the largest wind-energy facility in Canada. The wind farm is owned and operated by SaskPower the main electric utility for the province.

The wind farm has 83 turbines, each with a 1.8 MW capacity and a combined capacity of nearly 150 MW, which is enough to power about 65,000 homes. These sizes and capacities are typical for turbines and wind farms built in the first decade of this century. Turbine nacelles and blades are from Vestas, Denmark, but the towers were manufactured in Saskatoon by Hitachi (Fig. 7.20). All turbines are connected via underground cables to a central substation and transmission lines

Figure 7.20 Vestas V80-1.8 MW turbines positioned in wheat fields. Towers stand 67 m tall, rotor diameter spans 80 m, and total height reaches 107 m; total weight of each turbine is more than 200 MT (SaskPower 2024). The narrow access road and turbine pads are slight impediments for crop operations.

(Fig. 7.21). Electricity is sent via overhead 34.5 kV (kilovolt) transmission lines to the provincial grid system.

Turbines function at wind speeds between 15 and 90 km per hour (km/h), and shut off at lower or higher wind speeds. At hub height for this vicinity, average annual wind speed is 30 km/h, and full generating capacity is achieved at 50 km/h (SaskPower 2024). Given the cold winters of Saskatchewan, turbines are capable of running at temperatures down to −30 °C, but shut off at lower temperatures.

Turbines are scattered over an area of approximately 3,000 hectares in this rural windscape (SaskPower 2024). Individual turbines are positioned within agricultural fields, mainly cropland for wheat and canola, as well as cattle and horse pastures. The placement of turbines appears to have minimal impact on such rural land use (Fig. 7.22).

Figure 7.21 Electrical substation and transmission lines to left that serve the Centennial Wind Power Facility with a wheat field in the foreground.

Figure 7.22 Overview of turbines scattered across the agricultural landscape of the Centennial Wind Power Facility on the drainage divide between the Saskatchewan River and Missouri River basins. Kite aerial photograph (2024).

Figure 7.23 SaskPower, Coteau Creek Hydroelectric Station. Lake Diefenbaker on the South Saskatchewan River between Saskatoon and Regina.

7.5.3 Saskatchewan Energy Profile

Saskatchewan is a major oil-producing province, second to Alberta in Canada and fourth largest in North America after Texas and New Mexico (Statista Oil 2024). This includes conventional light oil as well as tight oil produced via steam–thermal means (Canada Energy 2024). Thermal bitumen projects and natural gas wells also add to fossil-fuel production. Coal has been mined since the 1850s and continues to be produced in large surface mines near Estevan in the southeast. Mining and extraction of fossil fuels, in fact, make up the largest segment of Saskatchewan's economy, accounting for more than one-quarter of the province's GDP (Statista Saskatchewan 2024).

Electricity generation in Saskatchewan comes mainly from natural gas and coal, about 40% each; the remainder is mostly from hydropower, and just a few percent come from other renewable sources, including wind (Canada Energy 2024). This balance of electricity generation is atypical for Canada (see above, Section 7.1), which may be explained by the abundance of local fossil fuels and hydropower (Fig. 7.23). The wind potential for southern Saskatchewan is substantial, but wind energy has far to go to make a sizable contribution to the province's mix of electricity generation.

7.6 Summary

Canada and the United States are the world's second and third largest countries, respectively, and the U.S. and Canada rank second and ninth, respectively, for installed wind-generating capacity. Both countries possess enormous fossil-fuel resources. At the same time, both countries are committed to increasing renewable energy and are poised for a major push to enhance wind energy.

Kansas has played a prominent role in the development of United States wind power and is currently ranked fourth among U.S. states for installed capacity. Nearly all wind farms in Kansas are situated on drainage divides in order to capture the best wind resource. Locations that are favorable tend to expand through time as additional turbines and wind farms are added. Furthermore, the size of turbines has increased through time in terms of height, rotor diameter, and rated capacity.

Colorado ranks seventh among states, and its wind farms are located almost entirely on the thinly settled High Plains east of the Rocky Mountains. As in Kansas, many of these wind farms are located on drainage divides. For both states, visibility is an issue given the height of turbines, some of which now exceed 180 m, and long sight lines to the distant horizon. For the most part, impact on wildlife appears to be minimal, and some sensitive areas are protected in various ways from future wind-energy development.

Offshore wind energy in New England has emerged as a critical component of the region's transition to a clean energy future. With strong and consistent wind resources, the region is positioned to lead the United States in offshore wind development. State-level policies have driven significant progress. However, challenges such as deep-water technology, environmental considerations, alignment of state and federal policies, and supply-chain constraints must be addressed. Continued collaboration between state governments, industry stakeholders, and research institutions will be essential in overcoming barriers and ensuring offshore wind energy plays a role in the region's renewable energy strategy.

The southern one-third of Saskatchewan is mostly open, agricultural land that is well suited for wind energy, particularly the Missouri Coteau, which forms the continental drainage divide between the Missouri–Mississippi and Saskatchewan–Hudson watersheds. Eleven wind farms have a total generating capacity of approximately 620 MW. Electricity generation in Saskatchewan comes mainly from natural gas and coal, and the remainder is mostly from hydropower. Just a few percent come from other renewable sources, including wind; thus, wind energy has far to go to make a sizable contribution to the province's mix of electricity generation.

References

Aber, J.S. and Aber, S.E.W. 2024. Kansas windscape: Wind energy in the 21st century. *The Kansas School Naturalist* 68(1), 13 p.

Aber, J.S., Aber, S.E.W. and Everhart, M.J. 2023. Roadside geology of Kansas. Mountain Press, Missoula, Montana, 307 p.

Aber, J.S., Aber, S.W. and Pavri, F. 2015. Windscapes: A global perspective on wind power. Multi-Science Publ., UK, 245 p.

Aber, J.S. and Ber, A. 2007. Glaciotectonism. *Developments in Quaternary Science* 6, p. 1–246.

Canada Energy 2024. Provincial and territorial energy profiles – Saskatchewan. Canada Energy Regulator. https://www.cer-rec.gc.ca/en/data-analysis/energy-markets/provincial-territorial-energy-profiles/provincial-territorial-energy-profiles-saskatchewan.html

Census Bureau 2022. QuickFacts: Huerfano County, Colorado. U.S. Census Bureau. https://www.census.gov/quickfacts/fact/table/huerfanocountycolorado/PST045222

CIA 2024. The world factbook. U.S. Central Intelligence Agency. https://www.cia.gov/the-world-factbook/

CoM 2022. An act driving clean energy and offshore wind. Bill H.5060. 194th General Court of the Commonwealth of Massachusetts. https://malegislature.gov/Laws/SessionLaws/Acts/2022/Chapter179

Davidson, R. 2018. Cape wind: Requiem for a dream. Windpower Monthly, May 1, 2018. https://www.windpowermonthly.com/article/1462962/cape-wind-requiem-dream

DECD 2023. Connecticut's offshore wind strategic roadmap. Department of Economic and Community Development. https://portal.ct.gov/offshorewind/-/media/offshorewind/state-of-connecticut-osw-strategic-roadmap-2023.pdf

Der, C., Horton, J. and Perry, K. 2024. Clean energy tax incentives in Budget 2024. Osler, April 18, 2024. https://www.osler.com/en/insights/updates/clean-energy-tax-incentives-in-budget-2024/

DOE Infrastructure 2022. Biden-Harris administration announces $30 million from bipartisan infrastructure law to speed up wind energy deployment. U.S. Department of Energy. [site removed by 2025 administration]

EIA Kansas. 2024. Kansas – state energy profile overview. U.S. Energy Information Administration. https://www.eia.gov/state/?sid=KS

Energy Atlas. 2025. All energy infrastructure and resources. U.S. Energy Atlas. https://atlas.eia.gov/apps/all-energy-infrastructure-and-resources/explore

EU 2021. What is the blue economy? European Commission, Oceans and fisheries. https://oceans-and-fisheries.ec.europa.eu/system/files/2021-03/what-is-the-blue-economy_en.pdf

Fitzpatrick, M. 2024. Offshore wind procurement. Research Report 2024-R-0200. Connecticut General Assembly Office of Legislative Research. https://www.cga.ct.gov/2024/rpt/pdf/2024-R-0200.pdf

GEO 2023. Maine offshore wind roadmap. Governor's Energy Office. https://www.maine.gov.energy/sites/maine.gov.energy/files/inline-files/Maine_Offshore_Wind_Roadmap_February_2023.pdf

Hartman, L. 2016. Computing America's offshore wind energy potential. Office of Energy Efficiency & Renewable Energy, U.S. Department of Energy. https://www.energy.gov/eere/articles/computing-americas-offshore-wind-energy-potential

Hoen, B.D., Diffendorfer, J.E., Rand, J.T., Kramer, L.A., Garrity, C.P. and Hunt, H.E. 2018. United States wind turbine database V8.0 (February 25, 2025). U.S. Geological Survey, American Clean Power Association, and Lawrence Berkeley National Laboratory data release. https://energy.usgs.gov/uswtdb/ or https://doi.org/10.5066/F7TX3DN0

Iberdrola 2025. Vineyard Wind I, our first offshore wind farm in the United States. https://www.iberdrola.com/about-us/what-we-do/offshore-wind-energy/vineyard-wind-offshore-wind-farm

IRENA. 2024. Renewable energy statistics 2024. International Renewable Energy Agency, Abu Dhabi. ISBN ISBN: 978-92-9260-614-5. https://www.irena.org/-/media/Files/IRENA/Agency/Publication/2024/Jul/IRENA_Renewable_Energy_Statistics_2024.pdf

Kansas Governor. 2020. Governor Kelly reaffirms Tallgrass Heartland wind moratorium. Kansas Office of the Governor. https://www.legistorm.com/stormfeed/view_rss/1905811/organization/101065/title/governor-kelly-reaffirms-tallgrass-heartland-wind-moratorium.html

MCC. 2025. Wind energy technology degrees and certificates. Mesaland Community College. https://www.mesalands.edu/academic-programs/degrees-certificates/wind-energy-technology-program/degrees-and-certificates/

McCoy, A., Musial, W., Hammond, R., Hernando, D.M., Duffy, P., Beiter, P., Pérez, P., Baranowski, R., Reber, G. and Spitsen, P. 2024. Offshore wind market report: 2024 Edition Golden, CO: National Renewable Energy Laboratory. NREL/TP-5000-90525. https://www.nrel.gov/docs/fy24osti/90525.pdf

Miller, P.R. 2023. Anti-wind politics could cost the Kansas economy. Emporia Gazette. https://www.emporiagazette.com/free/article_20871e36-8531-11ec-b27e-1745f79cec18.html

Musial, W., MacDonald, S., Fuchs, R., Zuckerman, G.R., Carron, S., Hall, M., Hernando, D.M., Sathish, S. and Fan, K. 2023. Considerations for floating wind energy development in the Gulf of Maine. Golden, CO: National Renewable Energy Laboratory. NREL/TP-5000-86550. https://www.nrel.gov/docs/fy23osti/86550.pdf

NEOW 2025. Massachusetts for offshore wind. New England for Offshore Wind. https://www.newenglandforoffshorewind.org/states/massachusetts/

NGMD 2023. The national geologic map database. U.S. Department of the Interior. https://ngmdb.usgs.gov/ngmdb/ngmdb_home.html

NOAA 2023. Information about offshore wind and whales. News, U.S. National Oceanographic and Atmospheric Administration. https://www.fisheries.noaa.gov/feature-story/information-about-offshore-wind-and-whales

Obermeyer, B., Manes, R., Kiesecker, J., Fargione, J. and Sochi, K. 2011. Development by design: Mitigating wind development's impacts on wildlife in Kansas. *PLoS One* 6(10), e26698. https://doi.org/10.1371/journal.pone.0026698

Office of Governor (Janet T. Mills) 2024. Governor Mills announces agreement on federal research lease to advance floating offshore wind. https://www.maine.gov/governor/mills/news/governor-mills-announces-agreement-federal-research-lease-advance-floating-offshore-wind-2024

Padbury, G.A. and Acton, D.F. 1994. Ecoregions of Saskatchewan. Saskatchewan Property Management Corporation, Central Survey and Mapping Agency, Regina, Canada.

Ropeik, A. 2024. The siting of an offshore wind port raises new conflicts in Maine. Grist. July 15, 2024. https://grist.org/energy/offshore-wind-port-siting-raises-new-conflicts-for-coastal-mainers-environmental-activists/

SaskPower 2024. Centennial wind power facility. https://www.saskpower.com/our-power-future/our-electricity/electrical-system/system-map/centennial-wind-power-facility

Sandercock, B.K. 2013. Environmental impacts of wind power development on the population biology of Greater Prairie-Chickens. DOE Scientific and Technical Information, Final Technical Report, DOE/EE0000526. http://www.osti.gov/bridge/product.biblio.jsp?osti_id=1080446

Shankman, S. 2025. Did just cancel offshore wind? Some projects off Massachusetts coast may be dead in the water: Making sense of the day one executive order that could change the energy game in New England. Boston Globe. https://www.bostonglobe.com/2025/01/21/science/trump-offshore-wind-executive-order/

SIEC 2014. Huerfano River wind project: 2014 wind cooperative of the year award entry, p. 9–23. San Isabel Electric Cooperative. U.S. Department of Energy. https://www.wapa.gov/wp-content/uploads/2023/04/SanIsabelElectAssoc14.pdf.

Singer, S. 2024. Feds reject Maine's request for $456 million to build proposed wind port at Sears Island. Portland Press Herald, October 22, 2024. https://www.pressherald.com/2024/10/22/feds-reject-maines-456-million-request-to-build-wind-port-at-sears-island/

Statista Oil 2024. Crude oil production in Canada in 2023, by province/territory; and crude oil production in the United States in 2023, by state. Statista. https://www.statista.com/statistics/484909/canadian-crude-oil-production-by-region/ and https://www.statista.com/statistics/714376/crude-oil-production-by-us-state/

Statista Saskatchewan 2024. Distribution of gross domestic product of Saskatchewan, Canada, in 2022, by industry. Statista. https://www.statista.com/statistics/608347/gdp-distribution-of-saskatchewan-canada-by-industry/

Stose, G.W. 1912. Description of the Apishapa Quadrangle, Vol. 186. U.S. Geological Survey, Geologic Atlas of the United States, Apishapa Folio. https://pubs.usgs.gov/gf/186/text.pdf

TNC 2022. Site renewables right: Accelerating a clean and green energy future. The Nature Conservancy. https://www.nature.org/en-us/what-we-do/our-priorities/tackle-climate-change/climate-change-stories/site-wind-right/??vu=siterenewablesright/?

TNC 2023. Flint Hills ranching to save the tallgrass prairie. The Nature Conservancy. Stories in Kansas. https://www.nature.org/en-us/about-us/where-we-work/united-states/kansas/stories-in-kansas/flint--hills-ranching-to-save-the-tallgrass-prairie/

UMASCC 2025. VolturnUS. University of Maine Advanced Structures & Composites Center. https://composites.umaine.edu/offshorewind/volturnus/

Ward, L.G., Bogonko, M. and Johnson, P. 2020. Northeastern U.S. Bathymetry and Backscatter compilation: Western Gulf of Maine, Southern New England and Long Island. University of New Hampshire Center for Coastal and Ocean Mapping and Joint Hydrographic Center, Durham. https://gis.ccom.unh.edu/portal/apps/experience/?builder/experience/??id=ae95dce059714ef2a70f052ecf4c4d96

White House IRA 2022. By the numbers: The Inflation Reduction Act. U.S. White House. [site removed by Trump administration]

Whitty, M. 2022. Assembly approves bill seeking major offshore wind energy development. Press Release, June 23, 2022. State of Rhode Island General Assembly News. https://www.rilegislature.gov/pressrelease/_layouts/RIL.PressRelease.ListStructure/Forms/DisplayForm.aspx?List=c8baae31%2D3c10%2D431c%2D8dcd%2D9dbbe21ce3e9&ID=372884&Web=2bab1515%2D0dcc%2D4176%2Da2f8%2D8d4beebdf488

Wind Power Canada 2024. Canada wind farms. The Wind Power: Wind energy market intelligence. https://www.thewindpower.net/country_maps_en_14_canada.php

WindExchange 2024. U.S. installed and potential wind power capacity and generation. WindExchange, U.S. Department of Energy. https://windexchange.energy.gov/maps-data/321

WindExchange 2025. Workforce training and education. U.S. Department of Energy, Office of Energy Efficiency & Renewable Energy. https://windexchange.energy.gov/training

WindExchange Offshore 2024. Offshore wind energy. Wind Energy Technologies Office, U.S. Department of Energy. https://windexchange.energy.gov/markets/offshore

WindExchange States 2024. Wind energy state information. WindExchange, Energy Efficiency & Renewable Energy, U.S. Department of Energy. https://windexchange.energy.gov/states/

8

Energy Production and Consumption

We want cheap energy where we need it, when we want it, and with no environmental impact (Tinker 2013, p. 8).

8.1 Overview

Homo sapiens are engaged in massive global changes through immense use of fossil fuels and other mineral resources, vast land conversions, and intense water consumption in order to support and raise the living standard for a rapidly growing human population, which now exceeds 8.1 billion people (USCB 2025). This undertaking is mostly uncontrolled; in other words, most decisions are made for immediate individual survival, gain, or comfort rather than the greater good of society or long-term environmental sustainability of the world.

Around the world, annual per-capita primary energy consumption varies tremendously. Among the various forms of energy, access to electricity is one measure of a country's general prosperity and quality of life. Iceland has by far the highest per-capita consumption of electricity in the world followed by Norway, Kuwait, and Bahrain (Table 8.1). Canada and the United States rank seventh and ninth, respectively, with annual usage of 13,900 and 11,700 kWh per person.

However, electricity consumption is certainly not the only indicator for a country's well-being. For example, Japan, Denmark, and the United Kingdom also have quite high living standards with much lower per-capita electricity consumption at 7,520, 5,630, and 4,700 kWh, respectively (Index Mundi 2020). At the other end of the spectrum, the world's poorest countries, such as Chad and Haiti, have less than 40 kWh per-capita electricity usage.

Wind Energy Landscape: Principles, Techniques, History, First Edition. James S. Aber, Susan E. W. Aber, and Firooza Pavri.

Published 2026 by John Wiley & Sons, Inc.

Table 8.1 Annual per-capita consumption of electricity for the top 10 countries as of 2020.

Country	kWh per person*
Iceland	50,400
Norway	22,400
Kuwait	19,300
Bahrain	17,400
Qatar	15,200
Finland	14,900
Canada	13,900
Sweden	13,100
United States	11,700
United Arab Emirates	11,300

* Rounded to nearest 100 kWh per person.
Adapted from Index Mundi (2020).

Increased generation and delivery of electricity are, thus, fundamental ingredients for economic development and human quality of life. The primary energy sources for generating electricity are fossil fuels (coal, oil, natural gas, peat), nuclear power, hydropower, and various forms of solar energy and wind. Other minor sources for electricity include geothermal power (Fig. 8.1), biofuels, and tides (see sect. 6.3.3).

Figure 8.1 Portion of a Japanese geothermal plant at Yamagawa that generates electricity from steam-powered turbines. This volcanically active area has high heat flow. Volcanoes result from subduction of the Philippine tectonic plate beneath the island of Kyushu. Since the earthquake and tsunami heavily damaged the Fukushima Daiichi nuclear plant in 2011, Japan has decided to place greater emphasis on solar, wind, and geothermal power.

Figure 8.2 (Left) Open-pit mega-mine in eastern Germany as seen in 1995. Miocene-age brown coal (lignite) was extracted for generating electricity and making concrete. (Right) concrete plant associated with German coal mine. Manufacturing concrete releases carbon dioxide from two sources, burning fossil fuel for thermal energy and the conversion of carbonate rock into lime (CaO) with the release of CO_2.

On a global basis, the total installed electrical-generating capacity is approximately 8,500 GW (CIA World 2025). Some 60% are produced by burning fossil fuels, 15% come from hydropower, and about 9% and 12% are generated from nuclear power and renewable sources (wind and solar), respectively.

Burning fossil fuels along with clearing forests, making concrete, and other human activities are the primary causes of increasing atmospheric carbon dioxide (Fig. 8.2). From a pre-industrial level of ~280 ppm, global mean atmospheric CO_2 concentration exceeded 420 ppm in 2023 and continues to increase steadily at roughly 2 ppm per year (NOAA 2025). Atmospheric CO_2 has already far surpassed the presumed tipping point of 350 ppm, beyond which uncertain or irreversible environmental consequences may take place (Foley 2010). As a greenhouse gas, CO_2 is intimately connected with the potential for global climate change. Beyond the enormous, but as yet uncertain, climatic implications for greenhouse warming, many other environmental, health, and aesthetic consequences result from mining and burning fossil fuels.

8.2 Fossil Fuels

Petroleum, coal, and natural gas are the big three fossil fuels. The International Energy Agency (IEA) has forecast that demand for fossil fuels will peak in the mid-2030s (Cuff 2022). Coal should decline first, followed by natural gas and oil. The Russian invasion of Ukraine has accelerated the trend toward renewable, nuclear, and low-carbon energy sources that will collectively reduce greenhouse-gas emissions.

8.2.1 Coal

In spite of the IEA prediction, coal provides more than one-third of global electricity generation, and demand for coal continues to grow. Major declines have taken place in the European Union, Australia, Japan, Korea, Canada, and United States (Fig. 8.3). In the latter, for instance, the number of coal-fired power plants has declined by more than half, and nuclear has fallen notably. Meanwhile, natural-gas power plants have increased by more than 20%, and renewable-energy power plants (solar and wind) have expanded phenomenally (Table 8.2). Meanwhile, the U.S. exports a substantial amount of coal to other countries (CIA United States 2025).

Figure 8.3 Trains pulled by diesel locomotives transport coal from mega-mines in Wyoming to power plants in the midwestern and eastern United States. BNSF Railway, Alliance, Nebraska.

Table 8.2 Change in U.S. electric industry power plants by main energy resources.

Year	Coal	Petroleum	Natural gases	Nuclear	Hydropower	Renewable*
2013	518	1,101	1,725	63	1,476	2,299
2023	227	1,075	2,084	54	1,481	7,684
% change	−56	−2	+21	−14	+0.3	+234

* Renewable includes wind plus solar power plants. Rounded values.
Adapted from EIA (2025).

The United Kingdom became the largest economy in 2024 to end the use of coal for generating electricity, although Belgium and Sweden had done so previously (Lawless 2024). The United Kingdom was the birthplace of coal power for the Industrial Revolution, and as recently as 1990, coal supplied about 80% of the country's electricity. The last coal-fired power station was closed in 2024 marking the end of an era. This was made possible by rapid growth of renewable sources (wind, solar, hydro, wave) and expansion of natural gas, a fuel that is much cleaner

Kansas Coal and Electricity

Kansas was formerly a major source for coal in the late nineteenth and early twentieth centuries, particularly from numerous coal beds in upper Pennsylvanian bedrock in the eastern portion of the state. Coal was used mainly for powering steam locomotives, smelting lead-and-zinc ore, and generating electricity. However, most coal mines ceased operating in the mid-twentieth century, and the last coal mine in Kansas shut down in 2016. Coal is still utilized for generating electricity at several power plants and accounts for approximately 38% of the state's electricity generation (EIA Kansas 2025). This is low-sulfur coal imported from Wyoming's Powder River Basin.

The recent role of coal in Kansas electricity generation is illustrated well by events at Holcomb, near Garden City. The current coal-fired electric-generating station at Holcomb was built in the early 1980s. It uses low-sulfur coal from Wyoming along with advanced technology to limit emissions of sulfur, nitrogen, and mercury gases. Its current-generating capacity is nearly 350 MW. This compares with more than 500 MW capacity for nearby wind farms—Buffalo Dunes and Cimarron I & II (Hoen et al. 2018–2025). A plan to expand the generating station was proposed in 2007.

The expansion proposal was rejected initially by the Kansas Department of Health and Environment, because of carbon dioxide emissions and concern about global warming. A revised and downsized plan for an 895-MW plant was finally approved in 2010, whereupon the Sierra Club filed suit to block the project. The state Supreme Court eventually approved expansion of the

Big Brutus, among the largest electric-power shovels in the world, operated in open-pit coal mines from 1963 to 1974. It was restored and opened as a public museum in 1985. Vertical view with silhouette shadow. Note smaller power shovel and bulldozer at bottom for scale. Near West Mineral, Kansas; helium-blimp aerial photograph.

(Continued)

Kansas Coal and Electricity (Continued)

generating station in 2017, but early in 2020, Sunflower Electric Power finally abandoned its effort to build the $2.2 billion coal-fired generating station after spending $100 million on the project (Aber and Aber 2020). Rapid growth of wind energy had made the coal-fired plant unnecessary; wind and solar energy produced about 40% of the state's electricity in 2025 (EIA Kansas 2025).

La Cygne Generating Station, a coal-fired plant located south of the Kansas City metropolitan area. In operation since the early 1970s with a capacity of 1578 MW. Subbituminous low-sulfur and low-ash coal is shipped by rail from the Powder River Basin in Wyoming. The plant is still in operation, but is planned for retirement during the 2030s decade (GEMW 2025).

than coal. Battery technology also plays a crucial role for this conversion (Poynting and Stallard 2024). The national goal is to generate all electricity from renewable sources by 2030.

On the other hand, coal usage has expanded significantly in eastern Asia—China, India, Indonesia, Viet Nam, and Philippines, which together consume more than 70% of global coal, and their increases more than offset the reductions in other countries (IEA Coal 2024). Coal is also the world's most carbon-intensive and dirtiest fossil fuel with serious environmental impacts and consequences for human health (Zakaria 2012).

As of 2019, worldwide emissions of carbon dioxide had exceeded 35 billion metric tons led by China, the United States, and India, which accounted for more than half the total (Table 8.3). Before assigning any blame for greenhouse gases, however, it would be wise to consider the question, who is responsible for carbon

Table 8.3 Top 10 countries for CO_2 emissions, given in billions of metric tons.

Country	CO_2	Country	CO_2
China	10.77	Germany	0.73
United States	5.14	South Korea	0.69
India	2.31	Iran	0.65
Russia	1.85	Canada	0.61
Japan	1.10	Saudi Arabia	0.58

Adapted from CIA Energy (2022).

emissions? Developed nations, in effect, have exported their dirty manufacturing, coal, and carbon emissions to China, which is now also by far the largest CO_2 emitter. Just as offshore banks are used to hide cash, developed countries employ offshore manufacturing to hide their true carbon usage.

8.2.2 Oil and Gas

Global shortages and supply interruptions for petroleum during the late twentieth century led to volatile prices, unstable energy sources, and chronic economic disruptions. The United States and many other countries faced energy insecurity. By the end of the twentieth century, the U.S. imported more than half of its petroleum. Petroleum imports peaked at 60% in 2005, and natural gas reached 18%; nearly one-third of total U.S. energy came from imported fuels (Beach 2013). Since then, the situation has changed dramatically as a result of new drilling and production techniques, namely directional/horizontal drilling and hydraulic fracturing, commonly known as fracking.

Fracturing is nearly as old as oil- and gas-well drilling itself. First tried in Pennsylvania in the 1860s, various explosive and acid techniques were utilized for many decades. The first modern *Hydrafrac* process was tested in Kansas in 1947 and subsequently developed in oil fields of Oklahoma and Texas (Montgomery and Smith 2010). Oil- and gas-bearing rock is fractured by fluid pumped into the well under high pressure, and the fractures are held open by some type of injected "proppant" such as sand, ceramics, or other granular materials.

The idea of directional drilling was suggested as early as 1873 in Pennsylvania (Sorkhabi 2013), and it was first tried in California in the 1930s in order to produce offshore oil from wells drilled on land. Since then directional techniques have been developed around the world for all types of oil- and gas-bearing formations. These

Figure 8.4 Oil-well drilling rig operating in the Uinta Basin of northeastern Utah, United States.

methods, used in combination, are necessarily complex and expensive to implement in practice; their usage spread gradually during the late twentieth century. The early twenty-first century has witnessed an oil-and-gas boom in North America (Fig. 8.4). In fact, the United States now exports considerable natural gas to other countries (CIA United States 2025). As a relatively clean fuel in abundant supply, electric utilities have turned increasingly to natural gas.

8.2.3 Greenhouse Gases

In spite of the positive impact of renewable energy and expanded use of natural gas, the atmospheric increases of carbon greenhouse gases (CO_2 & CH_4) show no indication of slowing and are predicted to continue an upward trend worldwide (Fig. 8.5).

The most optimistic and probably unrealistic model projects atmospheric CO_2 to stabilize ~450 ppm. Less optimistic forecasts show CO_2 concentrations rising to >500 ppm and beyond 800 ppm by the end of the twenty-first century. Any realistic hope to mitigate human impact globally on greenhouse gases means a combination of actually reducing the consumption of fossil fuels and effectively lowering carbon emissions from their use. Smil (2011) analyzed global energy usage and

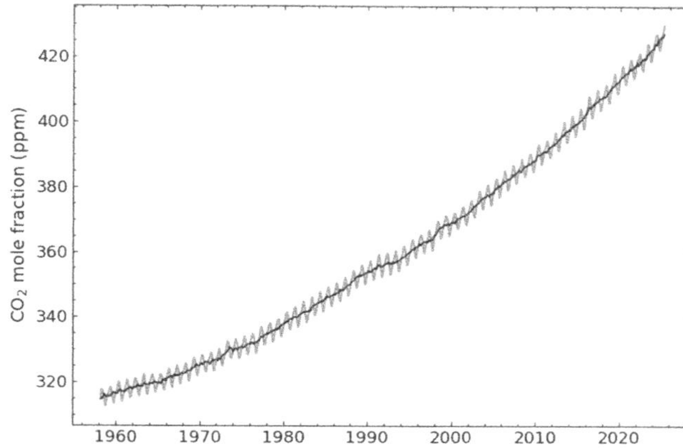

Figure 8.5 Chart of long-term trend in atmospheric CO_2 at Mauna Loa, Hawaii, through 2024. Saw-toothed line shows monthly variations; smooth line traces annual average. Atmospheric methane (CH_4) has a similar upward trend. NOAA (2025)/ National Oceanic & Atmospheric Administration.

technology. Comparing developed and developing countries, he reached two main conclusions.

- Most people in the world need to consume much more energy in order to enjoy reasonably comfortable, healthy and productive lives.
- The United States and Canada consume twice as much energy per capita as do the richest European Union countries or Japan, but do not have better standards of living.

He argued that per-capita energy consumption in the United States and Canada could be reduced substantially without a decline in quality of life, and this would promote innovation, strengthen the economy, and help the environment. He cited the serial failures of international agreements to reduce carbon emissions. He rejected numerous technological fixes to energy production and carbon pollution as logistically or financially unfeasible, and he emphasized the need for conservation and moderation of energy consumption. Many energy-saving methods have come to the fore in recent years, such as LED lighting, better-insulated and more-efficient buildings, mass-transit systems, and electric vehicles (Fig. 8.6).

8.3 Nuclear Energy

During the mid-twentieth century, nuclear energy experienced booming development with the promise of cheap, clean energy; reactors were built and uranium ore was mined in many locations (Fig. 8.7). Since then, nuclear energy suffered a series of spectacular accidents—Three Mile Island, Pennsylvania (1979), Chernobyl,

Figure 8.6 Typical fast-charging station for two electric vehicles (EVs). Such stations are becoming increasingly common in many countries. Seen here at Spodsbjerg ferry terminal, Langeland, Denmark.

Figure 8.7 Active, open-pit, uranium mine located in central Texas in the mid-1970s. Notice vehicles and people at bottom of pit for scale.

Ukraine (1986), and Fukushima, Japan (2011). From the late 1970s to the turn of the twenty-first century, the commercial nuclear industry suffered through a long recession, but it managed to maintain a significant share of global electricity generation.

Current global generation of electricity from nuclear energy amounts to about 2,600 billion kWh, which is slightly more than 10% of total electricity generation worldwide (Table 8.4; CIA Energy 2022). The top five countries—United States,

Table 8.4 Top 10 countries and world total for generation of electricity from nuclear energy as of 2023.

Country	Nuclear energy
United States	779
China, Mainland	406
France	324
Russia	204
South Korea	172
Canada	84
Japan	78
Spain	54
Ukraine	50
Sweden	47
World total	2,602

Values in billions of kWh.
Adapted from WNA (2024).

China, France, Russia, and South Korea—account for more than 70% of nuclear energy. The number of operating nuclear power plants peaked in the United States in the late twentieth century and has declined gradually since. However, upgrades at individual nuclear plants have yielded a high capacity factor overall, such that nuclear energy continues to supply nearly 20% of total electricity for the nation (EIA Nuclear 2023).

Only one new nuclear power plant has been built and completed in the United States during the current decade. Plant Vogtle in Georgia was proposed in 2008 and building began in 2013. It lagged years behind schedule, ran billions of dollars over budget, and was finally completed in 2023–2024 (Vogtle 2024). Delays are attributed to the 2011 Fukushima accident in Japan, 2017 bankruptcy of former contractor Westinghouse, and the coronavirus pandemic. Meanwhile similar delays and cost increases led to halting construction of another nuclear power plant in South Carolina in 2017.

Germany made the political decision to dismantle its nuclear power initially in 1998, and this plan was reintroduced in 2011 with the immediate shutdown of eight nuclear reactors. Prior to the shutdown, nuclear power provided about one-quarter of Germany's electricity; that figure is down to less than 2% now (WNA 2024). Sweden likewise has reduced its share of nuclear power considerably. On the other hand, some countries have increased nuclear power notably, such as

Belarus, Bulgaria, Finland, Pakistan, and Slovakia. Around the world, nuclear energy has grown by more than 10% during the past decade. Three reasons may be cited for this growth trend.

- Magnitude of electricity demand in developing countries.
- Recognition of energy security.
- Need to limit atmospheric carbon emissions.

The ultimate goal within the nuclear industry is reprocessing of all fuel using breeder reactors. However, this stage is decades away from commercial reality. In the meantime, increasing amounts of uranium ore must be mined and processed to supply the growing nuclear market. As with all mining, reclamation is necessary to mitigate long-term environmental consequences, and while this is mandated in most developed countries, it is not so likely to happen for many developing countries. The other serious concern is proliferation of nuclear knowledge, equipment, and fuel to potentially unstable countries. If the safety, mining, disposal, and proliferation issues could be managed, nuclear energy holds great promise for "green" electricity.

8.4 Wind Energy

Wind energy now exceeds 1,000 GW or about 6% of total installed electrical-generating capacity worldwide (CIA Energy 2022). Most wind energy currently is located in Asia, Europe, and North America, and these regions are likely to continue growing wind-power capacity rapidly in the near future. Wind generation of electricity is beginning to expand also in other regions, such as Australia and Brazil (see Table 1.3). The potential for wind energy, however, is certainly not uniform around the world.

Those regions with the greatest potential have mean annual wind speed at 50 m height >8 m/s. In North America, these include the U.S. Great Plains and southern Great Lakes, northern Canada and Canadian Prairie (Fig. 8.8), plus offshore Atlantic and Pacific coasts. Elsewhere, high wind potential is present in the British Isles and northwestern Europe, northern Siberia, desert regions of central and eastern Asia, northern Africa, western Australia, and southern South America, as well as many shallow coastal and oceanic locations (Aber et al. 2015).

Unlike fossil and nuclear fuels, the wind cannot be transported from its source to electrical power plants near consumers. In other words, much of the available wind resources exist in locales distant from where electricity consumption is most needed. Furthermore, wind-energy production is dictated by seasonal weather and diurnal patterns that may not be well suited for when people want to consume

Figure 8.8 Electrical substation and transmission line that serve the Blue Hill wind farm. This facility went online in 2022 with 35 turbines that have a rated capacity of 175 MW (SaskPower 2025). Located amid wheat and hay fields on the Missouri Coteau in the Canadian Prairie of southern Saskatchewan.

electricity. In the United States, high daily electricity consumption takes place between noon and midnight with peak loads around 6 pm and greatest demand in the summer. It is safe to assume that most people would not or could not change their daily and seasonal routines to coincide with available wind energy.

Energy storage has emerged as an important aspect for further harnessing wind power. Two forms of pumped storage have promise for application to wind energy (Rosen 2013).

- Pumped-water storage—water pumped from a lower source to a higher reservoir represents potential energy that may be converted into electricity when the water is released. Pumping takes place during periods of surplus power, and generation follows when extra electricity is needed (Fig. 8.9). Recovery of energy used for such pumping is typically 70–80% (see also sect. 6.3.3).
- Compressed air energy storage (CAES)—excess electricity is used to pump air into high-pressure underground formations near a power plant. Upon release, the air is heated and used to run conventional electric turbines. Man-made salt caverns have been used in Germany (since 1978) and Alabama (since 1991), and natural basalt formations have been tested in Washington (PNNL 2019).

Both pumped-water storage and CAES hold potential for integration with wind energy. The main drawbacks for both types are suitable topographic and geologic conditions in combination with existing electrical-generating capacity. For example, the Appalachian Plateau region of the northeastern United States would accommodate pumped-water storage, and artificial salt caverns in central Kansas could be utilized for CAES serving wind farms in the Great Plains region. Such favorable circumstances are not present, however, in all locales.

Figure 8.9 Pumped-water energy storage at Kinzua Dam and Allegheny Reservoir, Pennsylvania. The Seneca Power Reservoir (left) is situated about 210 m above the normal pool of the Allegheny Reservoir in the right background. Peak hydroelectric-generating capacity is 400 MW (USACE 2024). Kite aerial photograph (2009).

Battery energy storage systems, known as BESS, hold the best promise for widespread storage of excess wind and solar energy. In the United States, the Inflation Reduction Act (IRA) of 2022 provides for construction of 2,300 grid-scale battery plants (White House IRA 2022). A typical BESS has a capacity of 200 MW and up to 800 MWh, which is enough electricity to power 200,000 homes for four hours.

One concern with BESS is the risk of fire. Lithium-ion batteries are generally safe and are utilized in all types of devices from smart-phones to electric vehicles. But they have a known risk for fire and explosion when they are damaged or used improperly. Lithium-ion battery fires are especially dangerous, because water and fire extinguishers are not effective (FDNY 2025). This problem was highlighted in 2019 by a fire and unexpected explosion at a BESS in Arizona. Fire incidents have happened also in California, Idaho, and New York as well as Australia, France, and other countries (EPRI 2025). Still, fires at battery-storage facilities remain rare, and improved standards and procedures should minimize this risk.

Another geographic issue for wind energy is the electrical loss for long-distance transfer through existing power lines and grid systems. Developed countries typically have overall electrical losses of less than 10%. Total annual transmission loss in the United States, for example, is about 5% (USEIA 2023). Conversely most developing countries have higher losses, exceeding 20% in many cases (World Bank 2024). Adding wind energy from distant sources could increase potential losses.

Many aspects of electrical engineering influence the efficiency and cost of transmission lines, including voltage, conductor metal, wire construction and diameter, bundling multiple wires, and weather conditions (Harting 2010). In most situations, alternating current (AC) is favored over direct current (DC), because high-voltage AC can be stepped down to low-voltage distribution lines for delivery to consumers with minimal loss of power, but this is not feasible with DC.

Figure 8.10 Electrical substation and transmission line (on left) built to serve the Diamond Vista wind farm, which went online in 2018. Operated by Enel Green Power, near Tampa, Kansas, United States.

Primary loss comes from electrical resistance within the conductor as well as corona loss due to ionization of air molecules around the transmission line. For long-distance transmission, high-voltage lines are utilized, as loss is inversely proportional to voltage squared (V^2). For this reason, 765 kV transmission lines form the backbone of the United States grid system. Depending on details of line construction, 765 kV transmission typically experiences line loss of less than 1% over 100 miles (~160 km) distance (AEP 2011). A typical 345 kV transmission line, in contrast, suffers loss of more than 4%. Building such transmission lines is costly, in the range of several million US$ per mile (1.6 km) for a single-circuit 765 kV line through low-relief rural terrain. This explains why wind farms usually are sited near existing transmission lines. When wind-energy complexes are built or expand, electrical substations and high-voltage transmission lines must grow accordingly (Fig. 8.10).

8.5 Summary

Around the world, *Homo sapiens* are undertaking massive and mostly uncontrolled global changes in order to support and raise the living standard for a rapidly growing human population. Increased generation and delivery of electricity are fundamental ingredients for economic development and human quality of life. The primary sources for generating electricity are fossil fuels, nuclear power, and hydropower, as well as solar and wind energy.

Fossil fuels supply about 60% of electricity generation around the world. Among these fuels, coal provides one-third of global electricity and is the dirtiest fossil fuel with serious environmental impacts and consequences for human health.

Major declines have taken place in the European Union, Australia, Japan, Korea, Canada, and United States. However, demand for coal continues to grow in eastern Asia—China, India, Indonesia, Viet Nam, and Philippines, and these increases more than offset the reductions in other countries. Just as offshore banks are used to hide cash, developed countries employ offshore manufacturing to hide their true carbon usage.

Global shortages and supply interruptions for petroleum during the late twentieth century led to volatile prices, unstable energy sources, and chronic economic disruptions. Since then, the situation has changed dramatically as a result of widespread application of advanced drilling and production techniques. The early twenty-first century has witnessed an oil-and-gas boom in North America. As a relatively clean fuel in abundant supply, electric utilities have turned increasingly to natural gas. In spite of the positive impact of renewable energy and expanded use of natural gas, the atmospheric increases of carbon greenhouse gases (CO_2 & CH_4) show no indication of slowing and are predicted to continue an upward trend worldwide.

During the mid-twentieth century, nuclear energy experienced booming development with the promise of cheap, clean energy. Since then, nuclear energy suffered a series of spectacular accidents, but nonetheless has managed to maintain a significant share of global electricity generation. Some countries have reduced nuclear power considerably, while other nations are increasing their nuclear-power capabilities. However, the prospects for nuclear energy must deal with concerns about safety, mining, proliferation, high cost, and disposal of nuclear wastes.

Wind energy now exceeds 1,000 GW or about 6% of total installed electrical-generating capacity worldwide mainly in Asia, Europe, and North America as well as Australia and Brazil. Unlike fossil and nuclear fuels, the wind cannot be transported from its source to electrical power plants near consumers. Furthermore, wind-energy production is dictated by seasonal weather and diurnal patterns. This has led to several means for storing excess wind-generated electricity. Among these methods, battery energy storage systems (BESS) hold the best promise for widespread storage of excess wind and solar energy. Transmission of wind-generated electricity is another significant issue that requires construction of grid systems with minimal losses over long distances.

References

Aber, J.S. and Aber, S.E.W. 2020. Kansas wind power status. *Transactions of the Kansas Academy of Science* 123, p. 403–413.

Aber, J.S., Aber, S.W. and Pavri, F. 2015. Windscapes: A global perspective on wind power. Multi-Science Publ., UK, 245 p.

AEP 2011. Transmission facts. American Electric Power. https://web.ecs.baylor.edu/faculty/grady/_13_EE392J_2_Spring11_AEP_Transmission_Facts.pdf

Beach, F. 2013. Why is the U.S. so insecure about its energy security? *Earth* 58(5), p. 34–41. https://www.earthmagazine.org/article/why-us-so-insecure-about-its-energy-security-measures-energy-independence-show-it-increasing/

CIA Energy 2022. The energy world. U.S. Central Intelligence Agency. https://www.cia.gov/the-world-factbook/static/3a6e5fc200e47bc5cf68bd3a5d57481b/Energy_World.pdf

CIA United States 2025. CIA world factbook: United States. U.S. Central Intelligence Agency. https://www.cia.gov/the-world-factbook/countries/united-states/

CIA World 2025. CIA world factbook: World. U.S. Central Intelligence Agency. https://www.cia.gov/the-world-factbook/countries/world/

Cuff, M. 2022. Fossil fuel demand expected to peak within 15 years. New Scientist. https://www.newscientist.com/article/2344331-fossil-fuel-demand-expected-to-peak-within-15-years/

EIA 2025. Table 4.1. Count of electric power industry power plants, by sector, by predominant energy sources within plant, 2013 through 2023. U.S. Energy Information Agency. https://www.eia.gov/electricity/annual/html/epa_04_01.html

EIA Kansas 2025. Kansas: State profile and energy estimates. U.S. Energy Information Administration. https://www.eia.gov/state/?sid=KS#tabs-4

EIA Nuclear 2023. Nuclear explained: U.S. nuclear industry. U.S. Energy Information Administration. https://www.eia.gov/energyexplained/nuclear/us-nuclear-industry.php

EPRI 2025. BESS failure event database. Electric Power Research Institute. https://storagewiki.epri.com/index.php/BESS_Failure_Incident_Database

FDNY 2025. Dangers of lithium-ion batteries. Fire Department of New York City. https://www.nyc.gov/assets/fdny/downloads/pdf/codes/dangers-of-lithium-ion-batteries.pdf

Foley, J. 2010. Boundaries for a healthy planet. *Scientific American* 302(4), p. 54–57.

GEMW 2025. La Cygne generating station. Global Energy Monitoring Wiki. https://www.gem.wiki/La_Cygne_Generating_Station

Harting, C. 2010. AC transmission line losses. Stanford University, California. http://large.stanford.edu/courses/2010/ph240/harting1/

Hoen, B.D., Diffendorfer, J.E., Rand, J.T., Kramer, L.A., Garrity, C.P. and Hunt, H.E. 2018. United States Wind Turbine Database V8.0 (February 25, 2025). U.S. Geological Survey, American Clean Power Association, and Lawrence Berkeley National Laboratory data release. https://energy.usgs.gov/uswtdb/ or https://doi.org/10.5066/F7TX3DN0

IEA Coal 2024. Coal. International Energy Agency. https://www.iea.org/energy-system/fossil-fuels/coal

Index Mundi 2020. Electricity consumption per capita – World. Index Mundi. https://www.indexmundi.com/map/?v=81000

Lawless, J. 2024. Britain's last coal-fired electricity plant is closing. It ends 142 years of coal power in the UK. Associated Press. https://apnews.com/article/uk-last-coal-power-station-closing-9e6b792f27513e3b02904da1eaedfbf3

Montgomery, C.T. and Smith, M.B. 2010. Hydraulic fracturing: History of an enduring technology. *Society of Petroleum Engineers, Journal of Petroleum Technology* 62, p. 26–33. https://www.ourenergypolicy.org/wp-content/uploads/2013/07/Hydraulic.pdf

NOAA 2025. Trends in atmospheric carbon dioxide (CO_2). Global Monitoring Laboratory, U.S. National Oceanographic and Atmospheric Administration. https://gml.noaa.gov/ccgg/trends/

PNNL 2019. Compressed air energy storage. Pacific Northwest National Laboratory, U.S. Department of Energy. https://caes.pnnl.gov/

Poynting, M. and Stallard, E. 2024. UK to finish with coal power after 142 years. BBC. https://www.bbc.com/news/articles/c5y35qz73n8o

Rosen, J. 2013. Energy storage brings a renewable energy future one step closer. Earth, Vol. 58(10). http://www.earthmagazine.org/article/energy-storage-brings-renewable-energy-future-one-step-closer

SaskPower 2025. Blue Hill wind facility. SaskPower. https://www.saskpower.com/our-power-future/our-electricity/electrical-system/system-map/blue-hill-wind-facility

Smil, V. 2011. Global energy: The latest infatuations. *American Scientist* 99, p. 212–219. https://vaclavsmil.com/wp-content/uploads/2024/10/smil-article-2011-AMSCI.11.pdf

Sorkhabi, R. 2013. A letter from Pittsburgh: Reflections on the U.S. shale gas revolution. *Earth* 58(12), p. 37–38.

Tinker, S. 2013. Energy 360: Leaving our corners for the radical middle: New Zealand sets an example. *Earth* 58(9), p. 8–9. https://www.earthmagazine.org/article/energy-360-leaving-our-corners-radical-middle-new-zealand-sets-example

USACE 2024. Kinuza Dam and Allegheny Reservoir. U.S. Army Corps of Engineers. https://www.lrd.usace.army.mil/Missions/Projects/Article/3641480/kinzua-dam-allegheny-reservoir/

USCB 2025. U.S. and world population clock. United States Census Bureau. https://www.census.gov/popclock/

USEIA 2023. How much electricity is lost in transmission and distribution in the United States? U.S. Energy Information Administration. http://www.eia.gov/tools/faqs/faq.cfm?id=105&t=3

Vogtle 2024. Plant Vogtle is now the largest generator of clean energy in the U.S. Georgia Power. https://www.georgiapower.com/about/energy/plants/plant-vogtle/units-3-4.html

White House IRA 2022. By the numbers: The Inflation Reduction Act. U.S. White House. [site removed by the Trump administration]

WNA 2024. Nuclear generation by country. World Nuclear Association. https://world-nuclear.org/Information-Library/Facts-and-Figures/nuclear-generation-by-country

World Bank 2024. Electric power transmission and distribution losses (% of output). The World Bank. http://data.worldbank.org/indicator/EG.ELC.LOSS.ZS

Zakaria, F. 2012. The new oil and gas boom. *Time* 180(18), p. 20–21. https://time.com/archive/6596203/the-new-oil-and-gas-boom/

9

Ideal Energy

Bridging the energy gap will require a mix of resources, including reliable baseload power, intermittent resources, storage, and peaking plants (Elliott 2025, p. 9).

9.1 Radical Middle

Ideal energy sources would be available, affordable, reliable, and environmentally sustainable. These are the four major tenets for energy security (Tinker 2013). Wind energy is available in many regions, affordable and sustainable, but reliability is variable. Fossils fuels, on the other hand, are available everywhere and reliable. Affordability varies with geographic, political, and economic circumstances, however. The extraction of fossil fuels represents mining non-renewable resources. Furthermore, burning these fuels contributes to atmospheric pollution and potential global warming.

Nuclear power has great promise for green energy, but also has many intractable issues of safety, disposal of waste, high cost, proliferation, and public acceptance. Hydropower is clean and renewable, but large projects displace people, disrupt drainage systems, cut off downstream sediment, and impede migrating fishes. In other words, each major energy resource has notable strengths and weaknesses—no current or near-future energy source is ideal.

As Wald (2009) noted, *no technology provides a one-size-fits-all solution, but a combination can create a robust energy supply* (p. 58). Rather than taking sides in the energy debate, Tinker (2013) proposed the *radical middle* as the course to follow for analysis, decisions, and development of diverse energy resources. He emphasized the need to evaluate objectively the costs and benefits of fossil fuels,

Wind Energy Landscape: Principles, Techniques, History, First Edition. James S. Aber, Susan E. W. Aber, and Firooza Pavri.

Published 2026 by John Wiley & Sons, Inc.

including their environmental impacts, and to scrutinize the potential benefits of renewable energies likewise. This is not a simple undertaking as many costs are not obvious and possible impacts are uncertain.

9.2 Metals

Pursuing the radical-middle approach involves mining and processing ore, and manufacturing electrical components from several metals. Developing the infrastructure for wind energy especially requires more of certain metals, namely molybdenum, copper, and nickel, than for conventional generation of electricity based on fossil fuels (Matson 2013). In particular, nickel is necessary. Nickel is a key ingredient for casting steel-alloy components, such as rotor hubs, bearings, shafts, and gears that remain strong and ductile at low temperatures (Fig. 9.1; Coates 2021), and nickel–stainless steel and copper–nickel alloys provide corrosion resistance for turbines in the marine environment (Nickel Institute 2025).

Nickel comprises a portion of the Earth' core and is a component in nickel–iron meteorites (Fig. 9.2). In the Earth's crust, nickel is found in several geological settings and is mined at many sites around the world. Indonesia and the Philippines are the leading sources and account for more than half the world's production

Figure 9.1 Vestas V120-2.0 MW turbine in the Blazing Star I wind farm. Hoen et al. (2018–2025). Located on the Prairie Coteau of southwestern Minnesota. Low winter temperature in this vicinity is typically −10 to −15 °C and sometimes drops below −30 °C.

Figure 9.2 Nickel–iron meteorite from Muonianalusta, northern Sweden. The distinctive cross-hatched crystals represent the Widmanstätten pattern, which is made up by two nickel–iron minerals, kamacite and taenite that grew in an octahedral structure. This meteorite is the oldest known on Earth at 4.57 billion years (Blichert-Toft et al. 2010). Dime coin for scale (~18 mm).

(Statista 2025). Nickel ore comes mainly from two kinds of deposits (USGS nickel 2025).

- Laterites—formed by weathering of iron-rich igneous rocks. Principal ore minerals are nickel-limonite [(Fe,Ni)O(OH)] and garnierite (hydrous nickel silicate).
- Ultramafic rocks—deep-seated, iron-rich igneous rocks. Principal ore mineral is pentlandite, a nickel sulfide [$(Ni,Fe)_9S_8$].

Copper is among the earliest metals used by humans. It currently ranks third behind iron and aluminum for modern consumption, and it is essential for all types of electrical apparatus and infrastructure including generators, wiring, motors, and electronic devices (USGS copper 2025). Copper is found in various geological settings as a native element and is combined in many ore minerals—chalcopyrite [$CuFeS_2$], chalcocite [Cu_2S], cuprite [Cu_2O], and malachite [$Cu_2CO_3(OH)_2$], to name a few (Fig. 9.3).

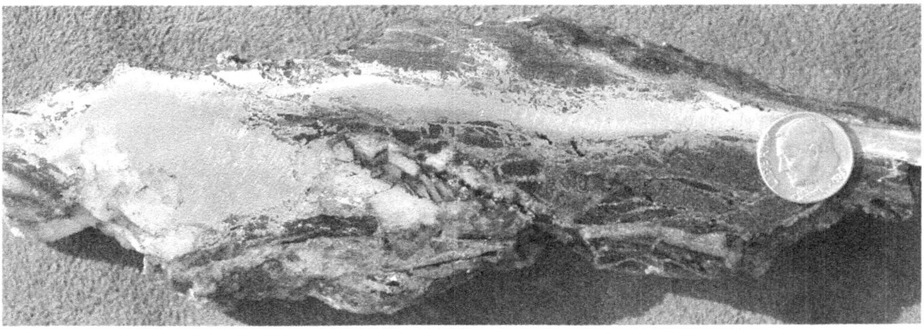

Figure 9.3 Native copper vein in a matrix of greenstone and quartz from northern Michigan, United States. Mining of native copper in this vicinity by Indigenous Peoples took place at least 8,000 years ago. USNPS (2025). Modern mining began in the 1840s, and most mining ended after World War II. Dime coin for scale (~18 mm).

Leading copper producers today are Chile, Democratic Republic of Congo, Peru, China, Indonesia, United States, and Russia (Nasdaq copper 2025). The forecast for copper trends expects volatility in the coming four years as global tariff turmoil increases (Belder 2025).

BESS facilities, electric vehicles, and many other devices depend on lithium-ion (Li-ion) batteries, which are the key to further electrifying the world's energy mix. Lithium, in fact, has become known as the "white gold" of the energy transition (LH 2024). Lithium is extracted from natural materials in two ways (Klein and Hurlbut 1999).

- Brine—concentrated saltwater solutions may contain high levels of dissolved lithium that are processed by evaporation and crystallization into lithium carbonate [$LiCO_3$].
- Minerals—Spodumene [$LiAlSi_2O_6$] and lepidolite [$K(Li,Al)_3(Si,Al)_4O_{10}(F,OH)_2$]. Both are found in granite pegmatites and produced by hard-rock mining and chemical treatments (Fig. 9.4).

Around the world, major lithium reserves are found in South America—Argentina, Bolivia, and Chile—Australia, China, and the United States. Australia, Chile, and China are the three main sources nowadays, which account for 90% of global lithium production (WEF 2023). Other metals necessary for Li-ion batteries include cobalt, nickel, manganese, copper, and aluminum.

Mining and refining these essential metals are energy intensive and involve significant carbon footprints and environmental disturbances (Fig. 9.5). In terms of the overall carbon budget, a new wind farm, BESS, and related infrastructure may take several years to break even with a conventional coal-fired power plant, which supports the argument to continue rapidly developing wind energy.

Figure 9.4 Lithium ore minerals found in pegmatites. (Left) Spodumene, variety kunzite (pink). (Right) lepidolite is a mica mineral that forms in thin sheets and typically displays pink, red, or purple colors. Dime coin for scale (~18 mm).

Figure 9.5 Bingham Canyon mine, Utah, United States, as seen in 2006. It is the largest human-made excavation and deepest open-pit mine in the world. The mine produces copper, gold, silver, molybdenum, and tellurium. The mine experienced massive landslides in 2013 and again in 2021, but continues in operation. These landslides demonstrated the vulnerability of such mining.

9.3 Wind Turbines and Health

Ideal energy should have minimal health risks. Rapid proliferation of wind farms has led to concerns that turbines may cause risk for human health, and much medical research has been conducted. Iowa is the third leading U.S. wind-energy state with 10 GW of installed capacity (see Table 7.1), and the state has a large a rural population living in proximity to wind farms. Thus, Iowa is a fair place to represent the potential health consequences of wind turbines. To summarize the possible health impacts of wind turbines, the University of Iowa College of Public Health undertook a comprehensive review of possible health issues.

Based on the best available research, the Iowa study concluded, *there is little scientific evidence that sound from wind turbines represents a risk to human health among neighboring residents.* Furthermore, *the only causal link that can be identified is that wind turbines may pose an annoyance to some who live near them. However, annoyance is likely influenced by a person's feelings about the impacts of wind turbines on viewsheds, whether they get an economic benefit from the turbines, whether they have had a say in the siting process, and attitudes about wind power generally* (Thorne et al. 2019). Similar results have been documented in many other locations around the world (e.g. McCunney et al. 2014).

Myths and Disinformation

Misinformation, distortions, and fake news have become increasingly common in recent years. Wind energy is not immune from this problem, which is driven by dubious news sources, social media, political agendas, and economic pressure. Some of these issues have been addressed in previous chapters:

- Wind-powered electricity is generated mostly at night—simply not true (sect. 3.1.2).
- Wind energy is unreliable—ignores regional nature of wind and grid systems (sect. 3.4).
- Texas energy fiasco—frozen wind turbines did not cause massive power failure (sect. 3.4).
- Wind turbines kill tens of millions of birds—not supported by research (sect. 5.2).
- Wind turbines are noisy—permitted noise equivalent to normal speaking voice (Table 5.1).

Many people involved with the fossil-fuel industry regard wind energy as an economic competitor. This point of view is questionable given rapid global increases in use of fossil fuels, which grew by nearly 50% between 2000 and 2023 (Ritchie and Rosado 2024). Nonetheless, fossil-fuel producers have waged a long-running and well-financed disinformation campaign against climate science and wind energy (Negin 2022; GWEC 2025). Various other disparate groups have joined the anti-wind movement as well. Furthermore, many large wind farms are owned and operated by major oil companies, which leads to a complicated situation (Westervelt 2024). More myths and disinformation are reviewed briefly.

- Anti-wind politics—conservative politicians across the United States are promoting legislation at the state level that would restrict development of wind farms (Miller 2023). They have branded wind energy as a "liberal symbol" that should be opposed on cultural rather than technical grounds. This point of view is divisive, shortsighted, self-serving, and cynical, given the large economic impact of wind energy.
- Offshore wind turbines kill whales—this myth started with whales offshore from the eastern United States and now has spread to other countries. However, scientific studies have not supported this claim. The primary cause of whale deaths is fishing gear entanglement, not wind turbines (Westervelt 2024).
- Wind turbines cause cancer—in spite of the repeated and unfounded claims by some politicians, absolutely no scientific evidence exists to link any form

Myths and Disinformation (Continued)

of human cancer with exposure to wind turbines. In fact, *greater exposure to wind turbines does not appear to increase cancer incidence* (Semprini 2019).

- Wind-turbine syndrome—many maladies supposedly are linked to wind turbines including vertigo, nausea, autism, tinnitus, fatigue, memory loss, depression, and migraines, to name a few. Taken together, wind-turbine syndrome is not a medically recognized diagnosis. According to the Ohio Department of Health, *there is **no** significant body of peer-reviewed, scientific evidence that clearly demonstrates a direct link between adverse physical health effects and exposures to noise (audible, LFN, or infrasound), visual phenomena (shadow flicker), or EMF associated with wind turbine projects* (ODH 2022).
- Wind-turbine radiation—exposure to electromagnetic field (EMF) radiated by turbines. Wind turbines do cause interference for AM radio reception. Some people have complained about this, but AM radio static is a common problem associated with high-voltage transmissions lines everywhere. According to the U.S. Environmental Protection Agency, *field measurements show EMF levels near wind turbines are well below any existing regulatory guidelines with respect to human health and are, in fact, lower than those produced by many common household electrical devices* (EPA Safety 2024).

Signs such as this often appear in the vicinity of proposed or new wind farms. Seen in Belleville, Kansas, approximately 10 km from the High Banks Wind Project that went online in 2023.

9.4 Energy in Transition

The switch from one primary energy source to another historically has taken several decades. Three such transformations have occurred during the Industrial Age, and the timing has varied from country to country (Smil 2014). The transition from wood to coal took place for some countries in the nineteenth century,

Figure 9.6 Open-pit coal mines in North America. (Left) Highvale coal mine in central Alberta as seen in 1984. Electric power shovels extract a thick seam of Cretaceous coal, which was burned in a nearby power plant to produce electricity. (Right) Big Brutus removes overburden from a thin seam of Pennsylvanian coal in southeastern Kansas in 1966. One of the largest electric-power shovels in the world, it stood about 50 m tall with a working weight of nearly 5,000 MT. All coal mines now have closed in Kansas, and Big Brutus is preserved as a museum (see sect. 8.2.1). Photograph courtesy of the Leslie and Beryl Ward collection.

although not until the mid-twentieth century for India and China. Coal remained the primary fuel throughout the twentieth century (Fig. 9.6), in spite of the dramatic growth of petroleum—the second transformation. A third transformation is underway now from coal and oil to natural gas, nuclear, and renewable resources—wind, solar, hydro, geothermal, wave, and tidal.

Some people regard wind energy and natural gas as competitors for generation of electricity, but they are not. In fact, they benefit each other because they compensate for one another. Wind energy is variable on a local basis, but this averages out over large regions, and cost is stable. On the other hand, natural gas is reliable and available, but the price is highly volatile. The price of natural gas in the United States, for example, fluctuated wildly between 2020 and 2024 from less than $2 per MMBtu (metric million British thermal units) to more than $9 per MMBtu. Price spikes were driven by the Texas energy crisis in 2021 and in 2022 presumably in response to Russia's invasion of Ukraine. Thus, natural gas and wind solve each other's reliability and price challenges (Webber 2012).

In the United States, in fact, electric utility companies are turning increasingly to a combination of natural gas, solar, and wind resources to provide reliable and affordable generation with reduced environmental impacts compared with coal and petroleum. As an example, Evergy is an electric utility that serves 1.7 million customers in western Missouri and eastern Kansas. Evergy plans to build two new, high-efficiency, natural-gas generating stations that would go online in 2029 and 2030. According to a company news release, *the two new plants will provide flexible generation that pairs well with the abundant renewable resource potential in*

Figure 9.7 OMV plant that processes so-called sour gas from the Schönkirchen and Höflein gas fields near Vienna, Austria. First developed in the 1960s and rebuilt most recently in 2016–2017. Notice numerous turbines of several wind farms in the background. C. Stadler/Bwag/Wikimedia Commons=/CC BY-SA 4.0.

Evergy's service area and will meet stringent emissions standards (Evergy 2024a). Evergy's goal is to become a carbon-neutral utility by 2045 with wind power as the largest single source for its generating capacity (Evergy 2024b). Similar developments are taking place around the world (Fig. 9.7).

For many countries, in fact, renewable energy already has become the primary source for generating electricity, for example, Albania, Angola, Brazil, Canada, Congo (DR), Costa Rica, Denmark, Ecuador, El Salvador, Ethiopia, Georgia, Iceland, Kenya, Laos, Lithuania, Luxembourg, Malawi, Mozambique, Nepal, New Zealand, Norway, Panama, Paraguay, Sierra Leone, Sweden (see Fig. 4.5), Tajikistan, Uganda, and Uruguay (CIA Electricity 2022). For the most part, these countries have favorable hydropower and/or geothermal resources. Denmark, Luxembourg, and Uruguay are notable exceptions, in which wind power is a major source for generating electricity. On the other hand, the world's largest countries—China, India, Russia, and the United States—continue to rely heavily on fossil fuels for generating electricity.

9.5 Energy Gap

Given this historical and current context, wind energy has a long way to go both for technical and financial reasons. Aside from Denmark and the United States, early advocates for wind energy (Fig. 9.8), most other countries did not start serious wind-energy developments until this century (Fig. 9.9). The difference

Figure 9.8 Early, relatively small Vestas turbine on the island of Fejø, southeastern Denmark, as seen in 1987. This turbine was erected by a local cooperative to serve the small island. Note access door and person for scale.

Figure 9.9 Vestas V80-1.8 MW turbines of the Centennial Wind Power Facility stand amid wheat fields of southwestern Saskatchewan, Canada. This wind farm was the largest in Canada when it went online in 2006.

between current energy production and future energy goals is the so-called energy gap (Elliott 2025). Major challenges confront the widespread expansion of wind energy (Smil 2014).

- Scale—fossil fuels supply more than 60% of global electricity generation (CIA Electricity 2022). For any new energy source to penetrate this huge market substantially is a formidable task given that world energy usage continues to increase rapidly (Ritchie and Rosado 2024).
- Integration—wind energy is intermittent in nature and often situated far from energy markets, which makes integration with existing supply systems difficult. Widespread adoption of wind energy would require substantial reshaping of the electricity infrastructure.
- Investment—the current energy infrastructure is vast, complex, and costly with payback times measured in decades. No country would quickly or easily abandon its existing energy facilities and investments.

Smil (2014) argued the best way to accelerate a transformation is to reduce overall energy consumption. With reduced demand, aging fossil-fuel sources may be retired and replaced by new wind, solar, and other renewable energy sources. This is a daunting task, as human population continues to grow with expectations for available, affordable, and reliable energy resources.

Dramatic expansion of wind- and solar-energy resources would be necessary to achieve complete renewable electricity generation by 2050. For wind power, this means significant growth for wind farms on land and offshore. Other renewable technologies—hydropower, biomass, geo- and solar-thermal, wave and tidal, or green-hydrogen systems—may contribute, *but no other generating technology will see anywhere near the growth that will be achieved through wind and solar technologies* (IRENA 2024, p. 12).

Developing safe and renewable energy resources that have minimal environmental impacts is a priority for humanity in the twenty-first century. Wind energy is part of the solution along with many other types and sources of energy. Diverse energy sources must be integrated into a production and delivery system in which the strengths of each type offset weaknesses for other types. Governments and society must look beyond narrow parochial, economic, and cultural self-interests toward long-term goals of human health, prosperity, and global sustainability.

9.6 Summary

Ideal energy sources would be available, affordable, reliable, and environmentally sustainable. However, no current energy source—fossil fuels, nuclear, hydro, wind, solar, wave-tide, geothermal—meets these criteria. No current or near-future technology is capable of providing a robust energy supply for a growing human population. The *radical middle* concept is an approach to follow for analysis, decisions, and development of various energy resources. This is a daunting undertaking as many costs are not obvious and possible impacts are uncertain.

Expansion of wind energy and further electrification of the world require more of certain metals, in particular aluminum, cobalt, copper, lithium, manganese, molybdenum, and nickel. Mining and refining these metals are energy intensive and involve significant carbon footprints and environmental disturbances. Disinformation and fake news have become increasingly common in recent years. Wind energy is not immune from this problem, but most of these myths are simply not true, including negative effects on human health.

A major transformation is underway now from coal and oil to natural gas, nuclear, and renewable resources—wind, solar, hydro, geothermal, etc. Some people regard wind energy and natural gas as competitors for generation of electricity, but they are not. In the United States, electric utility companies are turning increasingly to a combination of natural gas, solar, and wind resources to provide reliable and affordable generation with reduced environmental impacts, and similar shifts are taking place in many other countries.

Given the huge market size and enormous investments in fossil fuels, wind energy has a long way to go both for technical and financial reasons, which leaves a substantial energy gap. Dramatic expansion of wind- and solar-energy resources would be necessary to achieve complete renewable electricity generation by 2050. For wind power, this means significant growth for wind farms on land and offshore. This is a most difficult task, but developing safe and renewable energy resources that have minimal environmental impacts is a priority for humanity in the twenty-first century.

References

Belder, D. 2025. Copper price forecast: Top trends for Copper in 2025. https://www.nasdaq.com/articles/copper-price-forecast-top-trends-copper-2025

Blichert-Toft, J., Moynier, F., C.-T., Lee, A., Telouk, P. and Albarède, F. 2010. The early formation of the IVA iron meteorite parent body. *Earth and Planetary Science Letters* 296, p. 469–480. https://www.sciencedirect.com/science/article/abs/pii/S0012821X10003651

CIA Electricity 2022. Electricity generation sources. U.S. Central Intelligence Agency. https://www.cia.gov/the-world-factbook/field/electricity-generation-sources/

Coates, G. 2021. Wind and water – Nickel in clean energy. Nickel Magazine. Vol. 36(2). https://nickelinstitute.org/en/blog/2021/september/wind-and-water-nickel-in-clean-energy/

Elliot, P. 2025. Energy gap: Renewable paired with batteries. *Colorado Country Life* 56(1), p. 8–9. https://siea.com/uncategorized/energy-gap-renewable-paired-with-batteries/

EPA Safety 2024. Wind energy projects and safety. U.S. Environmental Protection Agency, WINDExchange. https://windexchange.energy.gov/projects/safety

Evergy 2024a. Evergy announces two new 705 MW high-efficiency natural gas plants. Evergy, News Releases. https://newsroom.evergy.com/2024-10-21-Evergy-announces-two-new-705-MW-high-efficiency-natural-gas-plants

Evergy 2024b. Our energy mix. Evergy, Renewable Resources. https://www.evergy.com/smart-energy/renewable-resources-link/our-energy-mix

GWEC 2025. Global wind report 2025. Global Wind Energy Council. https://www.gwec.net/reports/globalwindreport

Hoen, B.D., Diffendorfer, J.E., Rand, J.T., Kramer, L.A., Garrity, C.P. and Hunt, H.E. 2018. United States Wind Turbine Database V8.0 (February 25, 2025). U.S. Geological Survey, American Clean Power Association, and Lawrence Berkeley National Laboratory data release. https://energy.usgs.gov/uswtdb/ or https://doi.org/10.5066/F7TX3DN0

IRENA 2024. 100% renewable energy scenarios: Supporting ambitious policy targets. IRENA Coalition for Action. https://www.irena.org/-/media/Files/IRENA/Agency/Publication/2024/Mar/IRENA_Coalition_100_RE_scenarios_2024.pdf

Klein, C. and Hurlbut, C.S. Jr. (after J.S. Dana). 1999. Manual of mineralogy, 21st ed. Wiley, New York, p. 681.

LH 2024. Lithium extraction methods. Lithium Harvest. https://lithiumharvest.com/knowledge/lithium-extraction/lithium-extraction-methods/

Matson, J. 2013. Renewable energy's hidden costs. *Scientific American* 309(4), p. 100.

McCunney, R.J., Mundt, K.A., Colby, W.D., Dobie, R., Kaliski, K. and Blais, M. 2014. Wind turbines and health: A critical review of the scientific literature. *Journal of Occupational and Environmental Medicine* 56(11). https://doi.org/10.1097/JOM.0000000000000313

Miller, P.R. 2023. Anti-wind politics could cost the Kansas economy. Emporia Gazette. https://www.emporiagazette.com/free/article_20871e36-8531-11ec-b27e-1745f79cec18.html

Nasdaq copper 2025. Top 10 copper producers by country. Nasdaq. https://www.nasdaq.com/articles/top-10-copper-producers-country

Negin, E. 2022. Fossil fuel companies on the defensive. *Catalyst* 22, p. 8–11, 21. https://www.ucs.org/sites/default/files/2022-04/Catalyst-Winter-2022.pdf

Nickel Institute 2025. Nickel alloys in energy and power. https://nickelinstitute.org/en/nickel-applications/nickel-alloys-in-energy-and-power/

ODH 2022. Ohio Department of Health wind turbines and wind farms summary and assessments. Ohio Department of Health. https://odh.ohio.gov/wps/wcm/connect/gov/816f89dc-767f-4f08-8172-71c953b8ee02/ODH+Wind+Turbines+and+Farms+Summary+Assessment_2022.04.pdf?MOD=AJPERES

Ritchie, H. and Rosado, P. 2024. Fossil fuels. Our World in Data. https://ourworldindata.org/fossil-fuels

Semprini, J. 2019. Does exposure to wind turbines affect cancer incidence? A quasi-experimental analysis linking SEER and Geological Survey data in a hierarchical framework. Knowledge@UChicago. Open Access Repository for Documents, Data, and Media. https://knowledge.uchicago.edu/record/1927?ln=en

Smil, V. 2014. The long slow rise of solar and wind. *Scientific American* 310(1), p. 52–57.

Statista 2025. Distribution of mine production of nickel worldwide in 2022, by country. https://www.statista.com/statistics/603621/global-distribution-of-nickel-mine-production-by-select-country/

Thorne, P.S., Osterberg, D. and Johannsen, K. 2019. Wind turbines and health. Iowa Environmental Council. Environmental Health Sciences Research Center, University of Iowa College of Public Health. https://ehsrc.public-health.uiowa.edu/wp-content/uploads/2019/01/IEC_Wind_Health_Paper_2019_FINAL.pdf

Tinker, S. 2013. Energy 360: Leaving our corners for the radical middle: New Zealand sets an example. *Earth* 58(9), p. 8–9.

USGS copper 2025. Copper statistics and information. U.S Geological Survey, National Minerals Information Center. https://www.usgs.gov/centers/national-minerals-information-center/copper-statistics-and-information

USGS nickel 2025. Nickel statistics and information. U.S Geological Survey, National Minerals Information Center. https://www.usgs.gov/centers/national-minerals-information-center/nickel-statistics-and-information

USNPS 2025. Timeline of Michigan copper mining prehistory to 1850. U.S. National Park Service. https://www.nps.gov/kewe/learn/historyculture/copper-mining-timeline.htm

Wald, M.L. 2009. The power of renewables. *Scientific American* 300(3), p. 56–61.

Webber, M.E. 2012. Highlights of 2012: Outlook on natural gas. Earth: The Science Behind the Headlines. https://www.earthmagazine.org/article/highlights-2012-outlook-natural-gas/

WEF 2023. This chart shows which countries produce the most lithium. World Economic Forum. https://www.weforum.org/agenda/2023/01/chart-countries-produce-lithium-world/

Westervelt, A. 2024. The U.S. anti-wind movement, explained: Separating fact from fiction in the fight over offshore wind. Drilled. https://drilled.media/news/podcast-antiwind

10

Windscape Summary and Conclusion

10.1 Wind-Industry Origins

Electricity permeates all aspects of modern society in developed as well as developing countries around the world. This situation is hardly more than a century old; yet, electricity has become just as important as food, water, medicine, housing, and other basic necessities for economic prosperity and healthy life. Electricity is generated today from multiple renewable and non-renewable energy sources to meet the demands of a rapidly growing human population.

Half a century ago, wind energy was considered an unproven and generally unnecessary method for generating substantial electricity. In 1975, world population was just over four billion people, and electricity generation was 77 terawatt-hours (TWh). Population has doubled to more than eight billion now. During the same 50-year interval, electricity generation has expanded to ~185 TWh, a rate of increase greater than population growth (Ritchie et al. 2023; Ritchie et al. 2025; USCB 2025).

The unlikely rise of wind energy in the twenty-first century has deep historical roots beginning with traditional European windmills in the twelfth century (Fig. 10.1). This technology remained stable and functional into the early twentieth century. Meanwhile, American windmills were invented in the mid-1800s, proliferated around the world, and continue to be used widely (Fig. 10.2).

Modern wind turbines for generating electricity trace a direct line from Denmark in the 1890s led by Poul la Cour. His student, Johannes Juul, continued development of wind turbines in the mid-twentieth century, which became the so-called Danish concept. Rapid, innovative experiments and construction of wind turbines, large and small, took place in the 1970s by many inventors and self-builders. Fiberglass blades, pioneered by Ulrich Hütter in Germany, were adopted for Danish turbines.

Wind Energy Landscape: Principles, Techniques, History, First Edition. James S. Aber, Susan E. W. Aber, and Firooza Pavri.

Published 2026 by John Wiley & Sons, Inc.

Figure 10.1 Østofte Mølle, a traditional post mill located on the highest point of Falster island, southeastern Denmark. This mill was built in 1641 and continued in routine operation until 1943, more than three centuries. It was restored in 1970 and again in the 1990s, and is still in operating condition. The blocks in the ground are anchors for rotating the mill body around a central post so that the wheel faces into the wind (see also Fig. 1.2). Modern wind turbines are elaborated types of post mills, as the rotor and nacelle are perched on a tower.

Figure 10.2 Restored Dempster Vaneless No. 4 windmill from early twentieth century (1916 to mid-1920s) seen near La Veta, Colorado. The "horse" counterbalance arm points into the wind, and the blade sections are tilted in the downwind direction. Dempster was made in Beatrice, Nebraska, United States (Baker 1985).

By the early 1980s, the modern industrial-scale wind turbine emerged, which includes three fiberglass blades in a rotor mounted upwind from the tower, dual aerodynamic and mechanical braking systems, and asynchronous AC generator connected directly to the electricity grid (Fig. 10.3). This turbine setup and technology were exported widely in the late twentieth century to China, Germany, India, Spain, and the United States. Thus, the modern wind industry is essentially Danish in origin, and the small country of Denmark continues to play a leading role for further development of wind energy.

Figure 10.3 Micon wind turbine of 1990s vintage operating near Agerskov in southwestern Denmark, as seen in 2013. Note the tubular tower; rotor turns in clockwise direction as viewed from upwind (from right). This turbine embodies the Danish concept of the late twentieth century.

10.2 Windscape

Throughout this book, we have emphasized several wind-power themes—historical development, basic wind conditions, electricity principles, social and environmental issues, deployment in diverse countries, and its role within global energy usage and needs. The windscape represents the combination of these physical, technological, biological, and cultural elements for an integrated approach to assessing wind energy.

From this review, it is clear that development of wind energy has taken place in regions that have long histories of traditional windmill usage, particularly in Europe and North America. These places were culturally prepared to adopt newer and larger turbines for generating much-needed electricity. In other places, energy security and financial issues spurred development of wind energy along with concerns about greenhouse gases from burning fossil fuels and the potential for climate change. The world's three most-populated countries—China, India, and the United States—account for more than 60% of wind-energy capacity (IRENA 2024).

Nonetheless, uncontrolled development of wind farms took on a decidedly industrial appearance in some places with deleterious environmental impacts on flying wildlife—birds and bats, as well as nearby human residents. An environmental backlash began to develop. Siting and construction of wind farms came under stricter evaluation and control; new wind farms were prohibited in some locales, and older wind farms were required to repower in ways that reduced negative impacts.

Through time, wind turbines have become larger and larger with taller towers and bigger rotors in order to capture the best potential wind energy (Fig. 10.4). This, in turn, requires wider spacing between turbines on the ground and increasingly large land areas for new wind farms. Already in the 1990s, Denmark recognized the limits of land-based wind energy and, thus, began to move offshore. This trend accelerated in this century, particularly in China, Denmark, and the United Kingdom. Offshore provides greater wind potential with fewer environmental and human conflicts, although logistics and construction costs are greater than onshore. Offshore wind energy is nascent in France, Germany, Japan, Netherlands, Taiwan, and the United States (GWEC 2025), although India and Canada have yet to build any offshore wind farms.

Figure 10.4 GE Wind 2.8-MW turbines stand in agricultural fields within the Jayhawk Wind Farm of southeastern Kansas, United States. Individual turbines have a hub height of 114 m, rotor diameter of 127 m, and total height of 177.4 m with a blade in the upright position. Hoen et al. (2018–2025). This wind farm began operation in 2021, and these are among the tallest turbines in Kansas. Kite aerial photograph taken approximately at hub height (2022).

10.3 Wind and Other Energy Sources

Homo sapiens are undertaking massive and mostly uncontrolled global changes in order to support and raise the living standard for a rapidly growing human population. Increasing the generation of electricity is a key aspect, which presently depends heavily on burning fossil fuels—coal, petroleum, and natural gas. This contributes to steady rise in atmospheric carbon dioxide and other greenhouse gases that drive global climate change. Fossil-fuel and nuclear power plants represent so-called capacity resources that may be scheduled and adjusted as needed to meet demand (DOE Wind 2008).

Coal is the world's most carbon-intensive and dirtiest fossil fuel with serious environmental impacts and consequences for human health (Zakaria 2012). The use of coal for generating electricity has declined significantly in Europe and North America, but recent increases in eastern Asia more than offset those reductions (IEA Coal 2024). Natural-gas power plants have increased significantly, particularly in the United States, as this fuel is readily available and less carbon polluting, but its cost has fluctuated wildly in recent years. Nuclear energy holds great promise for "green" electricity, but continues to suffer from safety, mining, disposal, proliferation, and high-cost issues.

The case for wind energy is based mainly on its non-polluting, renewable, relatively low-cost characteristics. Furthermore, wind energy is produced locally and, thus, does not depend on imported fuels or unstable political conditions. The price structure for wind energy is likewise stable and not subject to unexpected fluctuations. Renewable energy resources, such as wind and solar, cannot be scheduled and adjusted as needed, however, as they depend on daily weather conditions. The irregularity of wind-generated electricity may be overcome in part with battery energy storage systems (BESS).

No single energy resource is ideal for generating electricity. Each has certain advantages, but also each has shortcomings or undesirable consequences. On this basis, Tinker (2013, 2014) proposed the *radical middle* as the course to follow for analysis, decisions, and development of diverse energy resources. In other words, using a mix of energy sources to generate electricity efficiently, minimize pollution and negative environmental consequences, and provide stable cost and price conditions. Full-cycle analysis extends from initial mining of metals and fossil fuels to eventual decommissioning of obsolete power plants.

A major transformation is underway now from coal and petroleum to natural gas, nuclear, and renewable resources. In the United States, electric utility companies are turning increasingly to a combination of natural gas, solar, and wind resources to provide reliable and affordable generation with reduced environmental impacts, and similar shifts are taking place in many other countries. Hybrid wind–solar–battery installations give the promise of enhancing the reliability of renewable energy.

10.4 Future of Wind Energy

Wind generation of electricity undoubtedly will continue to grow into the mid-twenty-first century with an emphasis on expanding existing favorable land-based vicinities as well as building more offshore sites. Local and national politics may play a significant role, however, in determining the rates and locations of wind-energy development. Some provinces and countries may move ahead aggressively deploying more wind-energy capacity, while others may become mired in bureaucratic inertia and political setbacks.

Myths and disinformation about turbines and wind generation of electricity have become commonplace and are increasingly accepted by the public and policy makers alike as reasons to oppose new developments. Indeed, opposition to wind energy has become a cultural issue to maintain the status quo. Fossil-fuel producers have waged a long-running and well-financed campaign against climate science and wind energy, in spite of clear evidence for the negative health and climate impacts of burning fossil fuels (Negin 2022; GWEC 2025). This point of view is divisive, shortsighted, self-serving, and cynical, given the large economic impact of wind energy.

Wind energy has a bright future. For those regions with adequate wind resources, human demand for electricity may be satisfied in part with sensible development of wind energy integrated with other energy sources and storage in well-designed grid systems. Wind energy is produced domestically and, thus, reduces the demand for foreign energy supplies and provides a modicum of energy independence for many nations. Lessons learned during the past half century may be applied forward to safeguarding environmental conditions and protecting aesthetic aspects of the windscape. Increased generation and delivery of electricity, including wind energy, are key ingredients for economic development and human quality of life.

References

Baker, T.L. 1985. A field guide to American windmills. University of Oklahoma Press, Norman, Oklahoma, 516 p.

DOE Wind 2008. 20% wind energy by 2030: Increasing wind energy's contribution to U.S. electricity supply. U.S. Department of Energy, DOE/GO-102008-2567. http://www.nrel.gov/docs/fy08osti/41869.pdf

GWEC 2025. Global wind report 2025. Global Wind Energy Council. https://www.gwec.net/reports/globalwindreport

Hoen, B.D., Diffendorfer, J.E., Rand, J.T., Kramer, L.A., Garrity, C.P. and Hunt, H.E. 2018. United States Wind Turbine Database V8.0 (February 25, 2025). U.S. Geological Survey, American Clean Power Association, and Lawrence Berkeley National Laboratory data release. https://energy.usgs.gov/uswtdb/ or https://doi.org/10.5066/F7TX3DN0

IEA Coal 2024. Coal. International Energy Agency. https://www.iea.org/energy-system/fossil-fuels/coal

IRENA 2024. Renewable energy statistics 2024. International Renewable Energy Agency. https://www.irena.org/-/media/Files/IRENA/Agency/Publication/2024/Jul/IRENA_Renewable_Energy_Statistics_2024.pdf

Negin, E. 2022. Fossil fuel companies on the defensive. *Catalyst* 22, p. 8–11, 21. https://www.ucs.org/sites/default/files/2022-04/Catalyst-Winter-2022.pdf

Ritchie, H., Rodés-Guirao, L., Mathieu, E., Gerber, M., Ortiz-Ospina, E., Hasell, J. and Roser, M. 2023. Population growth. Our World in Data. https://ourworldindata.org/population-growth

Ritchie, H., Rosado, P. and Roser, M. 2025. Energy production and consumption. Our World in Data. https://ourworldindata.org/energy-production-consumption

Tinker, S. 2013. Energy 360: Leaving our corners for the radical middle: New Zealand sets an example. *Earth* 58(9), p. 8–9.

Tinker, S.W. 2014. Energy 360: Moving energy – No easy feat, but a vital conversation. *Earth* 59(1), p. 8–9.

USCB 2025. U.S. and world population clock. United States Census Bureau. https://www.census.gov/popclock/

Zakaria, F. 2012. The new oil and gas boom. *Time* 180(18), p. 20–21. https://time.com/archive/6596203/the-new-oil-and-gas-boom/

Index

Wind Energy Landscape: Principles, Techniques, History, First Edition. James S. Aber, Susan E. W. Aber, and Firooza Pavri.

Published 2026 by John Wiley & Sons, Inc.

Printed and bound by CPI Group (UK) Ltd, Croydon, CR0 4YY
17/09/2026
14964311-0002